WESTERN TIMES

B R (W)

ISSUE No.7 - SUMMER 2023

Contents

The Transport Treasury
TIMES SERIES

Above: The station at Stourbridge Town looks very different today than it did in this 18 July 1959 view, and is now on a site some 70-yards further down the short branch towards Stourbridge Junction. Whilst still served by single unit railcars (in the form of the innovative Parry People Movers), all of the beautiful Great Western Railway infrastructure has long since been consigned to history under what is now the bus station. The ornate brick and stone station building was opened in October 1879 and survived for almost a century before being demolished in February 1979. The well maintained canopy, blue brick platform surface, gas lamp, original signage and a grounded Iron Mink van body provide a feast for modellers and historians alike. A pair of Tyseley allocated Class 122 Gloucester RC&W DMUs form the branch service, with the competing 'Midland Red' buses also in evidence. ***Douglas Twibell (1-25).***

Front Cover: Collett 2251 Class 0-6-0 No. 2218 stands at Brecon with the 12.15pm departure for Newport on Friday 10 July 1959. The June 1940 built locomotive was allocated to Newport Ebbw Junction (86A) for most of its working life, before a move onto the S&D at Templecombe for the final year in 1964. Ironically it was to return to Newport for scrapping at Cashmore's Yard in April 1965. For the technically minded, the photographer notes that the conditions on the day were 'Hazy Sun' leading to an exposure of 1/30 second at f.4 using Kodachrome film. ***Douglas Twibell (1-4).***

Rear Cover: To complement the Welsh flavour on the front cover, this is a Brecon & Merthyr Tydfil Junction Railway (B&MR) McKenzie & Holland lower quadrant somersault home signal. The location is approaching the former Church Road station on 13 October 1962. ***Douglas Twibell (45-23).***

ISBN 978-1-913251-45-1

First Published in 2023 by Transport Treasury Publishing Ltd.
16 Highworth Close, High Wycombe, HP13 7PJ.

Compiled and designed in the UK. Printed in Tarxien, Malta by Gutenberg Press Ltd.

INTRODUCTION

In the late summer of 2022 at the annual general meeting of a society of railway enthusiasts, the outgoing president revealed interesting information, some promising and some worrying, about prevailing trends in the hobby. This body had found it necessary to withdraw a well-patronised service for members following the retirement of the volunteer, unpaid manager and the society's inability to recruit a replacement. The hours that had been devoted over years by the retiree had demonstrated an extraordinary commitment to the role, and there was understandable reluctance for anyone else to take over an enjoyable job that nonetheless demanded punishing working hours and an onerous amount of travel. Engagement on commercial terms of an independent professional to run the service had been discounted on grounds of cost. This case graphically illustrated the degree to which the viability of the railway hobby in its many facets relies upon the often inadequately recognised commitment of voluntary labour to sustain the viability of many vital support activities.

Sitting at the rear of the meeting, physical evidence to support the president's key point lay before the observer in the almost 100% balding or silver-headed composition of the almost exclusively male audience. The president referred to the advanced average age of a membership that numbers well into four figures, and to the recurrent nature of this concern which has been prevalent for many years. Intriguingly, despite inevitable fluctuation in membership levels, the core contingent had remained essentially constant over that period, as had also the average age. Demographic attrition was naturally unavoidable but total numbers had been sustained by a steady supply of 'new' old men keen to join up. (Unfortunately no statistics were provided for female membership or for those who selflessly support their partners in the role of 'railway widow').

However, it is an ill wind and the impact of Covid that was considered to have had a positive influence on the manner in which this society had modified its operational style. A constituency which hitherto had lacked confidence in information technology has adopted Zoom meetings and embraced safe trading through the Internet. Psychological barriers persist but the reality that today's hobby increasingly exploits information technology as a legitimate part of the infrastructural framework augurs well.

The breadth and immediacy of modern communications produces a relentless flow of data drawn from sometimes scarce, almost priceless hard copy sources but the unprecedented scale risks submersion of important information in the dross. Production of Western Times, and doubtless its contemporaries, will continue to depend heavily upon the availability of traditional research methods. Digitisation of books, magazines, photographic collections etc is an invaluable aid to research but much remains to be uncovered through exploration of the dusty files and faded paper records of yesteryear. Translation into the repository of hard copy format is a service to those enthusiasts yet to be enlisted to the cause.

As diverse information provided through this publication accumulates, technology has been adopted to provide the means of tracking down topics appearing in earlier issues. Traditionally this would be provided intermittently by the distribution of a hard copy index, but this service is to be made available through an avenue of on-line enquiry. Details on how this service can be activated will appear in a future Guard's Compartment.

WESTERN TIMES SPECIALS

At our first editorial planning meeting, the idea was discussed of occasional 'special issues' that would focus solely on a particular theme. This concept took inspiration from the 1992 'Cornish Special' released under the *Great Western Railway Journal* label.

It is therefore with great pleasure that we can announce that the inaugural *Western Times Special* is scheduled for release this Autumn. This will feature an increased page count, but will retain the now established 'house style' and will of course continue to feature both black and white, and colour images. There could only be one subject worthy of our first publication, and that is to celebrate the centenary of the introduction of Charles Collett's iconic 'Castle Class'.

--- o O o---

Editor: Andrew Malthouse

Editorial Assistant: Jeremy Clements

To contact the editorial team please email: **WesternTimes@mail.com**

For sales and back issues of *Western Times* please go to: **www.ttpublishing.co.uk**

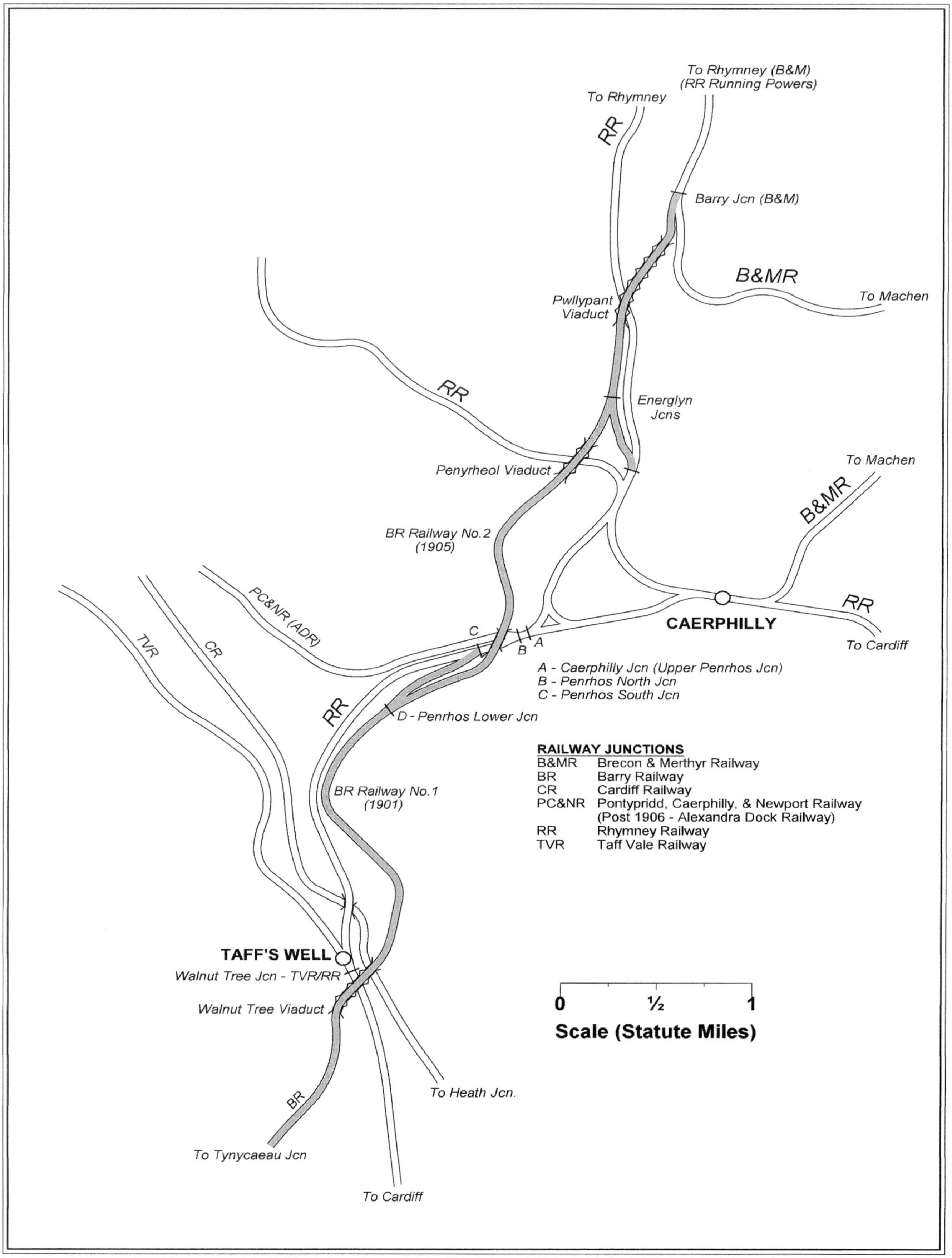
To Rhymney (B&M)
(RR Running Powers)
To Rhymney
RR
Barry Jcn (B&M)
B&MR
To Machen
Pwllypant
Viaduct
RR
Energlyn
Jcns
Penyrheol Viaduct
To Machen
B&MR
BR Railway No.2
(1905)
PC&NR (ADR)
TVR
CR
C
B
A
RR
CAERPHILLY
To Cardiff
A - Caerphilly Jcn (Upper Penrhos Jcn)
B - Penrhos North Jcn
C - Penrhos South Jcn
RR
D - Penrhos Lower Jcn
RAILWAY JUNCTIONS
B&MR Brecon & Merthyr Railway
BR Barry Railway
CR Cardiff Railway
PC&NR Pontypridd, Caerphilly, & Newport Railway
(Post 1906 - Alexandra Dock Railway)
RR Rhymney Railway
TVR Taff Vale Railway
BR Railway No.1
(1901)
TAFF'S WELL
Walnut Tree Jcn - TVR/RR
Walnut Tree Viaduct
0 ½ 1
Scale (Statute Miles)
BR
To Heath Jcn.
To Tynycaeau Jcn
To Cardiff

PENRHOS JUNCTIONS

TERRY McCARTHY

In *Western Times* Issue 2, a photograph appeared on page 39 showing No. 5653 on a train of coal empties. However, the caption was erroneous, so I wrote to the editor pointing this out (see 'The Guard's Compartment' Issue 5), whereupon a pleasant conversation ensued. Basically, it is wrong and certainly does R C Riley's knowledge of the area no favours. This concluded in an invite for me to write a correction of the caption, which has inevitably ended up becoming a more detailed overview of a complex location!

The original line was constructed and completed by the Rhymney Railway (RR) in 1858. For a short while this was the RR main line, extending from the RR junction with the Taff Vale Railway (TVR) in the Taff Valley at Walnut Tree, through the steeply graded cutting at Nantgarw into the Rhymney Valley at Aber (9.14 miles north of Cardiff) continuing to Rhymney itself. The line was built to serve the Iron Works at Rhymney and the slowly developing coal industry in the upper and lower parts of the Rhymney Valley.

In the history of railways in the Penrhos area there were four junctions to be noted, but they were built over a time span of 49 years by three different railway companies. These junctions are detailed below in date order and relate to the letters A-D on the map opposite:

A) The first junction opened in 1859, initially known as Caerphilly Junction and later as Upper Penrhos Junction, from about a quarter of a mile north of Penrhos, to serve two collieries at the southern end of Caerphilly. Later, in 1865, the branch linked with the Brecon and Merthyr Railway's (BMR) 3¾ mile-long single-track branch from Machen to an end-on junction with the RR a couple of hundred yards east of Caerphilly station. The branch was originally constructed in 1864 to link the BMR system to works and collieries. It remained a branch until 1871, by which time the RR, frustrated by the TVR's obstructionism and toll charges, completed construction of its own main route from Aber through Caerphilly to Cardiff, over part of the original RR (1859) Caerphilly branch. From 1871 the RR ran a shuttle service for passengers to the 'new' Caerphilly station from Walnut Tree Bridge station, just north of the junction with the TVR. After a few months, the service between Caerphilly-Walnut Tree was declared redundant and the section of RR line from Caerphilly junction to Caerphilly, first junction on the old 'main' line was closed in 1872. It was used thereafter for wagon storage, until 1884.

B) Junction two was located just north of Penrhos. It was constructed by the Pontypridd, Caerphilly and Newport Railway (PC&NR), as a result of the coal owners rebelling against the TVR and Cardiff Docks. Both seemed incapable of handling the traffic presented to them, especially after the Rhondda Valley started producing large quantities of good quality steam coal. Newport was seen as an alternative, hence development of the PC&NR which had close ties with Newport's Alexander Docks. Their route from Pontypridd to Newport, via Caerphilly and Machen was opened piecemeal in 1884/1886 and 1891. From 1884 the PC&NR linked its own line to the RR at a junction known as Penrhos North Junction, with running powers over the RR via the reactivated Caerphilly branch (closed in 1872) to the BMR branch in Caerphilly. From there, its coal trains (worked by the TVR until 1906 because of an agreement with the PC&NR) travelled over the BMR to Newport where the Alexander Docks Railway (ADR) took over to transport them into the docks. To handle the additional traffic the PC&NR built a loop (1891) between Caerphilly and Machen to avoid a steep gradient on the branch, effectively doubling the line.

C) The third junction, Penrhos South, was constructed by the Barry Railway (BR) in 1901, as part one of a three-stage project to extract coal from the Monmouthshire coalfield. Initially, traffic was mainly from the RR-served colliery at Treharris (owned by David Davies' Ocean Coal Company) which could be routed over the GWR and RR to the exchange sidings at Penrhos. Traffic from the London & North Western Railway (LNWR) Sirhowy lines could also be directed to Penrhos sidings, south of Penrhos South, helped by the GWR, having built a link with the RR at Hengoed (Ystrad Mynach North Junction – just over 9 miles north of Aber Junction), joining the RR main line in 1893 (from Rhymney to Aber Junction). From there trains continued down the RR main line (and a section of the original line) to Penrhos, whereupon this traffic was taken by the BR to Barry Docks. A proportion also went to Penarth Docks via Radyr. In 1912 745,000 tons of coal were transported to the Penrhos exchange sidings by the RR from these two main sources.

D) The Extension route chosen by the BR, was from Penrhos Lower Junction, the start of the next two stages of their proposed entry into the Monmouthshire Coalfield. This branched off from its own line, not the RR, thus technically it was not a Penrhos junction, but the name of the junction indicated the BR felt it was.

Opposite page: Map of Caerphilly's railways circa 1921. The highlighted line is that of the Barry Railway, which was built in two parts, as indicated. The Penrhos Junctions (ultimately four of them by 1905) are shown in the centre. ***Richard Harman.***

However, the line impacted on the Junction area when the BR-built viaduct over the RR and ADR lines opened in early 1905. This branch to Barry Junction BMR was 4¼ miles in length, cost about £300,000 (1905 prices!) much of which was spent building three viaducts, including that referred to above. Two others were constructed: at Penyrheol and Pwll-y-Pant (also known as Llanbradach, depending on where one lived!), the latter being 2,385 feet in length and almost 125 feet high over the Rhymney River.

The line opened officially on 2 January 1905, and via the BMR, served several collieries on the Monmouthshire side of the Valley, including McLaren Collieries numbers 1 and 3, New Tredegar, Elliot, Bargoed Pits and later Britannia. While no data has been found for traffic carried by the BR over this section of line, its significance to BR might be deduced from the 1918 Service Timetable, which showed that 28 loaded coal trains a day were scheduled to pass directly over the viaducts to Barry Docks of which 20 trains per day were scheduled to run from Penrhos Junction from the RR to Barry Docks. Nevertheless, the quantity of coal carried probably approached the 750,000 tons, predicted in the initial estimates of potential traffic given to the BR Board in 1901. The timetable also identifies the types of locomotives scheduled to work the route in 1918. Apart from one return working to McLaren No.1 colliery, worked by an L class 0-6-4T, the others were worked by B or B1 class 0-6-2T. Anecdotal evidence of workings by BR 0-8-0 locomotives (class D), has not been proved, although a photograph is said to exist (not seen by the author).

Below: The panoramic view from Walnut Tree Viaduct of Walnut Tree Junction (circa 1905) looking north, showing where the original RR diverged from the TVR main line. All signs of the RR station had been removed but the single-road RR locomotive shed (in the right foreground) remained, which it did until quite recently when it was demolished despite appeals for it to be listed. The shed was used by RR banking locomotives, but when the GWR took over in 1922, the shed was rationalised and the bankers reallocated to Radyr a couple of miles to the south. ***Terry McCarthy collection.***

Above: A comparative view of Walnut Tree Junction taken almost 50 years later, again captured from the lofty deck of the viaduct on 27 July 1952. The RR route to Penrhos can be seen diverging to the right immediately above the road bridge and the works of South Wales Forgemasters Limited now occupies most of the former yard site. Also of note is the newly opened (16 June 1952) spur crossing the fields above the station building, which connected the main line to the reopened Nantgawr Colliery and Coke Ovens. ***R C Riley (RCR 3808).***

A Bill for the extension into the Monmouthshire coalfield was presented to Parliament in 1905, but the opposition (mainly RR, GWR, and LNWR) was better prepared than previously. The BR had not, from its origins in the 1880s, been viewed favourably by the established railway and dock companies, particularly as it had made no secret of its desire to undercut the existing operators by syphoning coal traffic from their lines and charging lower rates for transport to Barry Docks. This maximised their income and shareholders' dividends, usually at the expense of the established facilities. It was noted too that their lines served few, if any, coal mines directly. Furthermore, finance from their rich and generally supportive shareholders who were more interested in profit than 'fairness', was available. With several MPs on their Board (it actively courted MPs) besides the members of the legal profession, BR and its enterprises had advantages the others lacked, probably engendering some jealousy. The company was also quick to go to law, despite the cost, consequently, existing operators found themselves having to fund prolonged, expensive legal battles. By the 1905 Bill, the opposition decided to use the similar weapons the BR employed, consequently, the 1905 Bill was rejected by Parliament.

During the railway's construction it was clear the BR wished to syphon coal from the Sirhowy and Ebbw Valleys, courtesy of its links with some of the coal owners in those areas. In response the BR made amendments to the proposed route (reducing the links with the BMR) presenting it in 1906 to Parliament, where it was again rejected. Further amendments led to a third Bill in 1907 which was superficially successful. However, Parliament added an additional 'clause' (Clause 23) which 'enacted that the rates per ton per mile to be charged in respect of the conveyance of goods and mineral traffic upon the railways by that Act shall be the same whether the traffic is destined for or coming from the harbour or docks at Cardiff, Penarth, Barry, or Newport'. Thus, it was a pyrrhic victory for the BR to which Mr. Lake, BR General Manager, suggested "it would be no use building the railway if this clause was operative." Legal battles continued at least until 1919, primarily for Clause 23 to be dropped or amended, often by slightly changing the intended route, notably by cutting out use of BMR tracks (in a Bill presented in 1912), a one-time ally. Nevertheless, the BR continued, unsuccessfully, to affirm its desire to build the lines.

Class 56xx on an empty minerals train heading up the first RR main line, towards Aber Junction sidings and the former route up the Rhymney Valley. By the date of this photograph the first main line had become a branch from Walnut Tree Junction. The single-track branch, built in about 1859 (original opening date not shown in minutes) gave access to collieries south of Caerphilly town. This line was closed in 1872, when a shuttle service from the RR station at Walnut Tree Junction (with the TVR) to the 'new' Caerphilly station (on the direct route to Cardiff) ceased. In 1884 the branch was reactivated the junction becoming known as Penrhos Junction, allowing coal trains from Pontypridd over the PC&NR and its trains of Rhondda and Aberdare coal (hauled by TVR locomotives until 1906). This line allowed PC&NR trains access to the BMR branch to Machen and thence to Newport Docks. When further junctions appeared after 1901, Penrhos Junction was renamed Penrhos North Junction. The RR line was doubled in stages between 1905-1929, while the PC&NR line was doubled in 1888. At the end of 1897 (some sources suggest 1906) the PC&NR became part of the Alexandra Dock (Newport and South Wales) Railway. This line finally closed in 1967. ***Courtesy the late Tom Sands.***

In the meantime, the Great War intervened and the BR (along with other south Wales Railways) was controlled by the Railway Executive Committee until 1921, when Parliament had decided to Group the railways into four large companies. The BR, reluctantly (they felt they had the financial muscle to go it alone), became part of an enlarged GWR in 1922, but economic conditions had significantly changed. The 'Coal Boom' was over, many local iron and steel industries declined and closed and the market for coal-firing ships, declined too. The railways also found themselves being challenged by road transport. Rationalisation became significant, having a profound effect on BR lines. The 1924 GWR Service Timetable shows only six trains a day worked each way over Llanbradach/Pwll-y-Pant Viaduct with another nine crossing Walnut Tree viaduct, demonstrating the decline in coal traffic. Traffic was now powered mainly by GWR locomotives including Class 56xx.

The GWR Directors' minutes in 1925 noted the intention of Cardiff Collieries (owners of Llanbradach colliery) to work coal beneath Llanbradach Viaduct. After discussion, the General Manager stated, "traffic could be dealt with by other routes". The Board agreed to apply for powers to abandon the Penrhos - Barry Junction route, which were granted in the GWR Act 1926, though the Board did not authorise abandonment of the line until 1929. Thus, after a short life this line was no more. Local observations suggest it was used for excursion traffic until the early 1930s (not confirmed), but in 1937 both the Penyrheol and Llanbradach/Pwll-y-Pant viaducts were demolished. Penrhos to Tynycaeau closed in 1967, with Walnut Tree viaduct being demolished in 1970. So ended the BR's Monmouthshire ambitions.

Today there are sadly no junctions remaining in the Penrhos area. Indeed the original line built by the RR was itself closed in 1982, as a direct result of the closure of many coal mines in south Wales and the cessation of the associated mineral traffic.

References:

Barrie, D S	*The Barry Railway*, Lingfield, 1962.
Bushby, J	*The Barry Railway Management circa 1906 &1907,* Welsh Railways Archive, Vol. VI no. 10 & Vol nos. 1 & 2, 2019/ 20.
Chapman, C	*Barry Railway Notes Vols. I – V,* Welsh Railways Research Circle.
Chapman, C	*Rhymney Railway Notes,* Welsh Railways Research Circle.
Cooke, R A	*Track Layout Diagrams of the GWR and BR (W.R.): Section 45 Caerphilly*, Lightmoor Press, 2022.
Kidner, R W	*The Rhymney Railway,* The Oakwood Press, 1995.
Page, James	*Forgotten Railways of South Wales,* David & Charles, 1979.
Periodicals	*GWR Magazine,* May 1937.
	Railway Magazine, June 1970.
	Railway Observer, Vol 36 March 1968.
Service timetables	*Barry Railway 1918.*
	GWR July-September 1924.
Newspapers	*Barry Herald, Weekly Mail, Western Mail, The Cardiff Times, Evening Express.*

Thanks also due to Didcot Railway Centre Archive, the National Library of Wales, Bargoed Library and Wikipedia (with the usual care).

The BR built its viaduct over the RR and ADR routes at Penrhos in 1905 as part of its scheme to tap coal produced in the Monmouthshire Sirhowy and Ebbw valleys. The GWR abandoned the route across the viaduct in 1929 following parliamentary approval (Act to abandon route passed in 1926) as there was adequate track capacity to ship coal from the Rhymney Valley by other lines. Looking west, Penrhos South Junction is in the immediate foreground while through the left-hand arch, the BR/ RR exchange sidings (shunted by RR locomotives) can be discerned in the background. These handled coal from the Rhymney Valley, notably Treharris Colliery. The BR loop line that linked with the BMR by way of the viaduct diverged at Penrhos Lower Junction, to the left at the far end of the exchange sidings out of sight, towards the upper far left. In this 1935 view, the approaching locomotive is Class 56xx, possibly No. 6616, with banking assistance probably provided by another 0-6-2T. The train comprises empty 20-ton 'Felix Pole' wagons hired to Stephenson Clarke who often supplied locomotive coal to other railway companies as well as shipping lines. The train is probably from Barry, bound for one of the RR-linked collieries. The target on the left-hand lamp iron appears to read B22 (B = Barry). ***Courtesy Llyfrgell Genedlaethal Cymru/ National Library of Wales.***

Above: During May 1964, Collett 0-6-2T No. 6635 of Radyr shed (88B) traverses Upper Penrhos Junction with a solitary LMS-design brake van from the Rhymney direction. The August 1928 built locomotive is seen carrying plain green livery and was withdrawn from traffic the following month. The lines diverging to the right are to Caerphilly, hence the original name of Caerphilly Junction. *John Wiltshire, courtesy Peter Brabham.*

Two images taken showing the trackbeds of the various lines in 1989 from similar angles to those on pages 8 & 9. In the right hand view, whilst the brick viaduct pillars remain, the RR junction has disappeared under vegetation, and the PC&NR line, while visible is rapidly being taken over by undergrowth. Today the entire Penrhos cutting is filled and covered by a housing estate. *Terry McCarthy.*

CAERPHILLY CASTLE - 100 NOT OUT

It will surely not have gone unnoticed amongst all devotees of the Great Western Railway, that 11 August 1923 saw the official release into service of No. 4073 *Caerphilly Castle*. The pioneer of C B Collett's seminal four-cylinder express passenger 4-6-0 design, following construction at Swindon Works was first allocated to Old Oak Common, where following inspection by the GWR Directors began a working life of 36 years and 9 months. Withdrawal came on 10 May 1960 and after cosmetic restoration, preservation at The Science Museum in South Kensington and latterly in STEAM - The Museum of the Great Western Railway in Swindon. This short pictorial tribute to the iconic centenarian will be augmented with a *Western Times Special* issue later in the year.

As built outside Swindon Works in August 1923, with original bogie brakes, cab rainstrip and 3500 gallon tender. ***GWR Official.***

Locomotive and tender en-route from Park Royal to the Science Museum for preservation, 4 June 1961. ***R C Riley (RCR 15861).***

THE VERSATILE COLLETT 2251s

FROM THE ARCHIVES OF R C RILEY

The last pre-Grouping 0-6-0 produced by the GWR was No. 381 in July 1902, through the simple expedient of converting the 'Sir Daniel' Class 2-2-2 of the same number. Twenty-three members of the class had been so treated when Churchward halted the programme on taking over at Swindon. Other than locomotives absorbed at the Grouping, no more examples of a wheelbase that had been a mainstay of freight traffic in the previous century were added until 1930. By then, the 0-6-0 Dean Goods had been in service since 1883 and the older examples were falling due for withdrawal.

Collett with characteristic minimalism introduced a replacement whose chassis was dimensionally similar to the earlier class in its final form (cylinders, wheelbase, wheel diameter). The superheated tapered Standard No. 10 boiler had appeared about 5 years earlier for use mainly with ex-Taff Vale 0-6-2Ts and ex-Midland & South Western Junction 0-6-0s. Many of the tenders were second-hand and were the main source of visual variation. The only notable innovation was the full width, side-window cab which evoked interest among observers and appreciation among footplate crews used to the *al fresco* décor of older 0-6-0s. They sometimes referred to the class as 'little Castles'

Known as Class 2251 after the prototype's number, there were eventually 120 in service. Construction took 17 years 10 months to complete with Nos. 3218/ 9, the only 0-6-0s introduced to service by British Railways. Their status meant that in service they rarely attracted much attention but nonetheless undertook a wide range of duties. Personal recollections include their adding a 'touch of class' to the final stages of the Somerset & Dorset Joint and the memorable sight, having just alighted from the Exmouth ferry at Starcross in about 1957/ 8, of a 2251 leading a 4-6-0 County on an Up boat train comprising 16 bogies. The pair were working hard and presumably the 0-6-0 had been an emergency replacement at Newton Abbot. Fortunately, Dick Riley was well aware of their presence across the network.

The class featured quite prominently among his pre-war shed scenes as with No. 2252 at Swindon on 15 August 1938. The second to be built and completed in March 1930, here it was obviously fresh from a works overhaul. Early examples were equipped with lever (pole) reverse but information on how many plus their specific identities is hard to track down. An official photograph of No. 2251 shows a horizontal reversing rod emerging from the cab front slightly above splasher height and curving down beneath the frames, just in front of and following the arc of the centre splasher. Apparently the deluxe version with screw reverse was introduced quite early and those with pole reverse might have been modified to conform, although this is not certain. Nos. 2251-70 were fitted with single horizontal handrail below the cab side window but several including this locomotive later had a second vertical rail added ahead of the window. Early class members were recipients of second-hand tenders as with this 3500-gallon example that was significantly older than its companion. Withdrawals started in 1959 and No. 2252 was taken out of service in December of that year. ***R C Riley (RCR 275).***

Right: Another early Dick Riley image of compromised quality, but historically important nontheless. The date and location of this pre-war view of No. 2253 are unrecorded in the archive but judging by the sparkling condition, this is another that was recently out of the works. The mixed traffic mainline nature of its intended duties is apparent in the steam heating hose and the ATC equipment. Another aged tender bearing the 'shirt button' emblem has been attached. It seems to have been loaded with coal but looks light on its springs. The cabside has the original arrangement of horizontal handrail only; the chimney seems to be copper-capped which must have been quite unusual for a secondary locomotive type in the 1930s. This engine was a late survivor, being withdrawn on 8 March 1965 from Worcester shed (85A). *(RCR 220).*

Below: No. 2286 was at Worcester on 12 March 1939. Those numbered 2271-90 were built with the additional vertical handrail on the cabside. The story of the GWR's association with war-surplus Railway Operating Division 2-8-0s derived from Great Central Class '8K' was complex but by 1925, the company had completed purchase of 100 examples. Fifty judged suitable for short-term service only were withdrawn between 1927 and 1931 while their tenders were retained for further use. Equipped with vacuum brakes, many were paired with Aberdare 2-6-0s. Having been built for service in continental Europe, these tenders lacked water pick-up equipment but their 4000-gallon and 7-ton carrying capacities made them useful on longer distance freight duties. Withdrawal of 'Aberdares' commenced in 1934 and some of the ex-ROD tenders were then passed to Class 2251. Nos. 2281-6 were paired with them from new and at least another eight worked with them at some stage during their careers. No. 2286 built in January 1936 apparently retains the ex-ROD tender that it acquired when built. This engine was withdrawn in September 1964 and some of the ex-ROD tenders survived until around that time. *(RCR 159).*

WESTERN TIMES

Opposite page top: On 28 May 1951, Machynlleth-allocated No. 2201 is seen emerging in a southerly direction from the Fairbourne cliff shelter on the Cambrian coast line. This 56-metre long reinforced concrete structure was built by the GWR in the mid-1930s as a belated measure to protect the track from landslip and falling rocks. On two previous occasions in January 1883 and again in March 1933, locomotives had been derailed and sent plummeting to the shoreline below after striking debris at this exact location.

Over three years into nationalisation, the grimy locomotive has a modern 3500-gallon tender but has yet to receive its smokebox number plate. The first vehicle in the train is a Southern Region General Utility Van followed by a Toplight but it is impossible to determine whether it is a passenger brake van or a brake third. Work was found for many of the class until quite late in the steam era on the Western Region; this locomotive survived until June 1964 and was scrapped at Swindon in January 1965. *(RCR 3296}.*

Opposite page bottom: Further evidence of the class's role in Wales is found with No. 3207 also of Machynlleth shed standing at the level crossing at Barmouth station on 26 July 1951 with a passenger service. The gates have been closed in favour of the train, so departure must be imminent despite the crew being distracted by events on the promenade and beach! Plain black livery with the early BR crest suited the proportions of the 2251s, although strangely none retained this colour with the later post-1956 emblem. No. 3207 was part of Lot. 360 and emerged from Swindon in late October 1946, spending most of its working life at Welsh sheds, apart from two years at Worcester prior to withdrawal in December 1962. Following time in store at Worcester Works, it was to meet its demise at the hands of the cutters torch in Cashmore's yard at Great Bridge in July 1964. *(RCR 3275).*

Below: Resumption of construction in January/ February 1938 with Nos. 2292/ 2291 respectively saw final resolution of the cabside handrail issue with a single L-shaped handrail that bordered the window profile except that in this case there was no window. All those in service at the outbreak of war (Nos. 2200-10/ 51-99) had their windows removed and metal plates affixed over the opening as a blackout precaution. Those built during the war (Nos. 2211-40) were built with solid cabsides (i.e. no opening) but had L-shaped handrails adjacent to the area where the window would otherwise have been. Post-war cab modification and insertion of side windows took some years to complete, again reflecting a low order of priority. No. 2238, standing at the familiar location of Machynlleth shed yard on 25 July 1951, had yet to be modified. (Sister No. 2239 still had window-less cabsides in May 1955).

Starting with No. 2241, most locomotives entered service with the modern form of 3500-gallon tender with continuous fender, and as a result of rotation, No 2238 has acquired an example in lined black livery with 'BRITISH RAILWAYS' in plain font. There are signs of similar lining on the cabside beneath the grime, which interestingly extends up around the missing window aperture. Another of the class is standing behind while in front there is an ROD tender suggesting the presence of a third. By then the class was common on services in Central Wales. *(RCR 3269).*

MONA CAFE
BRITISH RAILWAYS
3207

Above: Machynlleth-allocated Nos. 3200 & 3201 were running into Bala Junction on Saturday 28 July 1951 with an inter-regional service comprising ex-LMS stock of differing ages. The train has the appearance of being a Summer extra from somewhere east of Wrexham bound for the Welsh Coast, possibly Pwllheli by way of Dolgelley and Barmouth. The train is running into the loop line platform more normally used by Ffestiniog branch services. Bala was a comparatively modest junction but it bears the GWR trademark for such a location – plenty of signals.

Some of the class survived into the last year of WR steam; No. 3200 was withdrawn in January 1965 and its companion the following May. Although part of the post-war build, both are coupled to older tenders. *(RCR 3292).*

Left: In England, the class was useful on main line intermediate passenger stopping services as with No. 2247 of Worcester shed climbing Hatton Bank with a Leamington-Worcester service on 24 September 1955. The train comprises what seems to be three non-corridor coaches; a Collett Composite, a Collett Brake Third and finally a Hawksworth All Third. The locomotive was in service until February 1964. *(RCR 6893).*

Hatton Bank again on 20 April 1957 with Tyseley's No. 2279 tackling the climb on the relief line. According to the BR headcode, this is a pick-up goods working and all the open wagons in sight are apparently empty. This engine was one of the first withdrawals in January 1959. *(RCR 10453)*.

Above: Although usually associated with rural pursuits, the class also found work at built-up locations. On 12 October 1957, No. 2222 was passing Old Oak Common East Signal Box at the head of what appears to be a transfer freight working. *(RCR 11328).*

Below: Some class members participated in the lined green 'Indian Summer' livery of the late 1950s. Secondary locomotives supposedly devoted to passenger services were so decorated although these engines were sometimes called 'Collett Goods'. On 7 April 1958, No. 2207's lined green can be detected beneath the grime. The location is Gloucester, its home shed and it seems to be awaiting repairs. The chalked message on the smokebox side states 'TUBES TO RUN (?) BEFORE GOING INTO SERVICE'. The cabside insignia shows that the class had a B power rating and yellow route availability whereas predecessor Dean Goods were uncoloured. *(RCR 11553).*

Evidently working as station pilot, locally allocated No. 2249 was shunting carriage stock alongside Ayleston Hill Signal Box at Hereford Barrs Court on 18 August 1962. The coaches are a Diagram D127 Collett Brake Third (possibly No. 670), probably a Diagram E160 Composite and furthest away, a Hawksworth D131/ D133 Brake Third. The Collett coaches were culled in 1964/ 5 while the Hawksworth vehicle survived a little longer. The locomotive was taken out of service in September 1964. *(RCR 16731)*.

POUND GREEN SIGNAL BOX

A study of the excellent records available from the Signalling Record Society (and previously available through the late George Pryer) reveals the existence of a break-section signal box at Pound Green on the Basingstoke branch south of Reading, four miles north of Mortimer and just over two miles south of Southcote Junction.

Under normal circumstances a six mile block section might not have been considered excessive for the train service on the line, notwithstanding of course that the route was an important cross-country link to the LSWR at Basingstoke and in consequence there were regular through workings.

World War 1 made those lines that reached north-south more important. Troop and supply movement to the south coast ports became paramount and in consequence two break-section signal boxes were provided on the Basingstoke branch. The first was at Stratfield Saye, between Mortimer and Bramley which opened on 11 May 1917 and the second that now under discussion, at Pound Green.

Pound Green opened on an unknown date in 1917 and besides breaking up the previous six mile block, also controlled access and egress to a down loop. Aside from that, we know the box itself was on the west side of the line, with the loop on the opposite side of the running lines. Pound Green remained in use until both and it and the loop were decommissioned after just nine years in September 1926. The wooden structure was dismantled and subsequently reused at Firestone on the Brentford branch where it survived until 31 May 1964.

Move forward in time 14 years from 1926, and again line occupancy was at a premium and once more in time of conflict. A new Pound Green signal box and loop was commissioned on 16 February 1940, the box once more of timber, but this time to an unknown design as records are scant. This second box had a life of 10 years surviving until 27 February 1950.

A signalling diagram survives for the 1940 installation but not for that of 1917 although realistically they were probably similar. In both eras a closing switch was provided but there is a difference in the type of lever frames installed and the number of levers. In 1917 it was 18 levers at 5¼ inch centres, in 1940 16 levers at 4 inch centres. What this means is that possibly a second hand frame was used in 1917 and this had extra levers but which were not used. An alternative might have been the 1917 layout including a crossover between the up and down lines so as to allow an up service to be shunted clear if required.

The two photographs accompanying this short piece were obtained over 30 years ago when interviewing the family of former railwayman Jim Squelch, whose father-in-law (name not recalled) started his signalling career at Pound Green in 1917 and who is depicted in the images. It was recalled his mother would often bring him food when he was working the box hence the likely source of the images. Unfortunately, no images have been found of the 1940-1950 box.

Further down the line, Stratfield Saye followed a similar pattern, the first box closing in 1924 and a second box open between 1910 and 1950. Both the original and later structures here were much smaller affairs with 4 and 6 levers respectively.

With grateful thanks to Roger Simmonds.

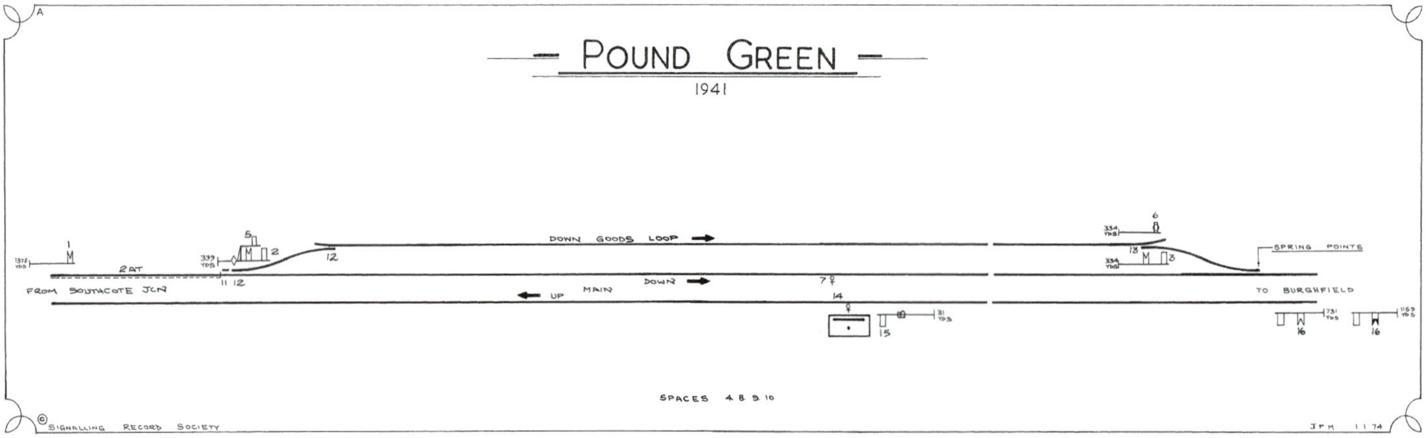

Right and below: Local snapshots of the first box at Pound Green from about 1925, prior to its relocation to Firestone. Moving signal boxes was not all that unusual. Sometimes structures were no longer needed at their original location or required to be moved to a new site at the same location in consequence of track layout alterations.

Wooden boxes of this type were made in sections and whilst the roof may well have had to be rebuilt, the walls and floor could be dis-assembled and transported by rail to the new site. A simple break-section box such as this was also an ideal starting point for a career in signalling, although it is likely there would still have been some time with a broom on the platform first! Dependent upon traffic, it is possible the signalman here may also have been responsible for the lamps which were likely to have needed refilling and trimming every 24hrs.
Kevin Robertson collection.

MSWJR LOCOMOTIVES UNDER THE GWR

MIKE BARNSLEY

BACKGROUND

After the Railways Act of November 1921 decreed that the railways should be grouped into just four large Companies, it was determined that the Midland & South Western Junction Railway (MSWJR) would be absorbed by the Great Western Railway. In anticipation of the takeover, a dossier was prepared of the MSWJR equipment. This showed that the MSWJR then had 29 locomotives divided into eight different classes. In readiness for the takeover, the GWR allocated spare numbers from their lists. These ranged from 23 to 1336. When the GWR took control, the renumbered locos were provided with cast GWR numberplates, which differed from the standard plates by having 'GWR' above the number.

The GWR actually took control of the MSWJR on 29 October 1923. The GWR will have known that the MSWJR, with its own Works at Cirencester, had tried to keep its locos in good order, but that the heavy military traffic during the First World War had stretched the Company's resources, resulting in a backlog in repairs and overhauls. When the GWR took control there were six locos needing overhaul or repair.

GWR numberplate from absorbed ex-MSWJR Dubs 2-4-0 No. 11.

DUBS 4-4-0 No.9

No.9 was the oldest loco in the MSWJR fleet, dating from 1893, and had been parked behind Swindon Town shed in need of a major overhaul. The GWR had little use for a small express 4-4-0, and she was condemned on 21st January 1924, without ever carrying her GWR number 1127.

MSWJR 4-4-0 No.9 at Swindon Town shed around 1920.

DUBS 2-4-0 Nos. 10 - 12

An unidentified MSWJR 2-4-0 at Swindon Town station around 1920.

Nos. 10-12 were the next oldest in the MSWJR fleet, dating from 1894. Nos. 11 & 12 were the first two MSWJR locos to be sent to Swindon in November 1923. No. 12 had been laid aside at Cirencester Works with boiler problems, and No.11 was due for overhaul. No.10 spent the November at Andover shed, and worked 1,657 more miles before joining her sisters at Swindon. All three were then rebuilt with GWR Standard No.11 Belpaire boilers, a saturated type introduced in 1924 for fitting to absorbed locomotives. They were also given new cabs and renumbered 1334–1336, but were never superheated. No. 12 re-entered service as 1336 in July 1924, and went to Swindon GW shed. Nos. 1334 & 1335 followed in the September, but went to Didcot, where No. 1334 was to remain for most of her time until 1952, apart from brief visits to Fairford, Lambourn, Newbury and Reading. In 1925 No. 1335 spent brief periods at Fairford and Lambourn, and in the summer of 1926 spent some time at Winchester before returning to Didcot. In 1925 No. 1336 spent a few weeks back on home territory at Andover Junction before moving to Oxford. During the 1930s she occasionally spent brief periods at Lambourn, and in July 1931 was briefly at Wallingford. In December 1935 she moved to Reading, but was back at Didcot in March 1936, eventually transferring to Reading in July 1938 where she stayed until 1952. In February 1926 No. 1336 moved to Reading, where she remained until 1952.

As small mixed traffic locos, the GWR made good use of them on the Lambourn Valley line, working both passenger and freight trains. In between trips to Lambourn, they served as station pilot at Newbury, occasionally being called upon to work specials to Oxford. In 1937, No. 1336 was noted regularly working the 8.15 a.m. Newbury – Reading fast passenger, which entitled it to the express headcode, which was quite an achievement for a 43-year old 2-4-0! The return working was the 4.38 p.m. to Marlborough, briefly visiting her old MSWJR territory but being a stopping-train, it did not qualify as an express.

The trio were the last MSWJR locos to remain in service, not being laid aside until March 1952. All three were then put into store at Swindon, with No. 1334 being scrapped in the September. However, both Nos. 1335 and 1336 had a swansong. In August 1952 No. 1335 was taken out of store and sent to Oxford shed to work a Stephenson Locomotive Society special to Shipston-on-Stour on 31 August. The *Railway Magazine* reported that she was then seen at work in Oxford, Reading and Slough in the early part of September, but she returned to Swindon towards the end of the month and was promptly sent for scrap like her sister. This left just No. 1336 in store at Swindon but in June 1953 she was revived to run a Gloucestershire Railway Society special over the MSWJR, and was not scrapped until 1954.

Ex-MSWJR 2-4-0 No. 1335 is seen taking water at Oxford shed, whilst working the SLS special to Shipston-on-Stour on Sunday 31 August 1952. ***R C Riley (RCR 3811).***

The un-rebuilt loco was shown on GWR 2-4-0 diagram A7, the rebuilt locos on diagram A9. The power classification was initially ungrouped, but was later altered to 'A'. In their final years they were put into British Railways power class '1'. The route availability was uncoloured.

DUBS 0-6-0T Nos. 13 & 14

Nos. 13 & 14 were also built in 1894. When new they were mostly used on freight services, but in later years were employed on the Andover-Tidworth and Chiseldon-Swindon local passenger services. As a result they were kept in good condition and allowed to keep their crimson lake livery, when freight locos had generally been repainted plain black. The GWR saw no reason to make any changes apart from fitting new numbers 825 and 843; even the MSWJR lamp sockets were retained. When due for overhaul in 1926, further cost was considered unjustified and both were sent for scrap.

They were shown on GWR 0-6-0T diagram B16. Their power rating was ungrouped and route availability uncoloured.

MSWJR 0-6-0T No. 14 at Swindon after receiving GWR numberplate 843.

BEYER PEACOCK 0-4-4T No. 15

MSWJR 0-4-4T No.15 at Andover Junction, around 1921. ***A W Croughton.***

Beyer Peacock No.15 was built in 1895 at the Gorton Works in Manchester, to the same drawings as those used previously for a locomotive supplied to the Wirral Railway. She was intended for local passenger services. Allocated to Swindon Town MSWJR shed when the GWR took control, she remained there until June 1924, when taken into Swindon Works and rebuilt. But while a photograph taken in March 1923 shows her in original condition, others taken at Swindon GW shed in May 1924 show her already modified with a GWR cone over the safety valves and bunker sides extended upwards, although still in MSWJR red livery complete with monogram and the original number 15.

The subsequent rebuild was to be more drastic, as the original round-top boiler was replaced with a GWR Belpaire boiler, the bunker was enlarged and she was given a new cab and the number 23. She was not superheated. Upon returning to traffic in June 1925 she was sent to Swindon GWR shed, and from April 1927 spent most of the time working the Swindon Junction – Swindon Town shuttle (often referred to as the 'Old Town Bunk') until December 1929, when taken into the Works and promptly condemned. The unrebuilt loco was shown on GWR diagram 'F', the rebuilt loco on diagram 'G'. Power rating was ungrouped, and route availability colour yellow.

Ex-MSWJR 0-4-4T No.23 at Swindon Junction station, c1928, almost certainly on the 'Old Town Bunk' to Swindon Town.

BEYER PEACOCK 2-6-0 No. 16

In 1892 the MSWJR obtained permission from the LSWR to work freight trains into Southampton Docks. The Company therefore began looking for a locomotive to handle such traffic. It was essential that she should be powerful, so that they didn't have to turn work away to avoid overloading. Speed was not a priority. In 1895 the Company therefore purchased No. 14 from Beyer Peacock, who built her to the same drawings as a class of 2-6-0 locos they had previously supplied to the New South Wales Government Railways in Australia. The MSWJR were so pleased with their purchase that they ordered a second loco, with detail differences, which became their No. 16.

With outside cylinders and 4-foot driving wheels, the pair gave their crews a lively ride and weren't favoured by the Company Engineer as they were hard on the track. As a result, when No. 14 developed frame problems in 1913, she was taken out of service and her boiler removed and installed at Cheltenham shed to provide hot water for washing out. The frames were sold for scrap, but were eventually rescued to provide a locomotive for the Ministry of Munitions during the First World War. No. 16 remained at work throughout the war, acquiring the nickname 'Galloping Alice' on account of her lumbering gait (see *Western Times* Issue 4, p.36).

MSWJR 2-6-0 No.16 at Swindon Town shed in 1921. ***A W Croughton.***

By the time of the GWR absorption, No. 16 was out of use awaiting overhaul and was taken into Swindon Works in December 1923. Despite the unusual design she was rebuilt with a superheated Standard No. 9 boiler and a new cab. Renumbered No. 24, the locomotive was also given a GWR tender, which did not line up with the loco running plate. She was returned to traffic at Swindon GW shed in February 1925, only to be put in the Stock Shed in March and taken back in to the Works in June. Her rapid return to the Works was presumably for rebalancing in an effort to give her a smoother ride, as when placed back into service in November 1925, there were much larger balance weights on the driving wheels. She was then put to work on the Swindon – Stoke Gifford (now Bristol Parkway) freights, a task she performed regularly until stopped in May 1930 and finally condemned in the July.

Before she went, No. 24 had a moment of glory. She was shunting at Badminton when 'King' class No. 6003 *King George IV* failed there with a broken whistle stem while hauling the 11.45 am Bristol – Paddington express. Being vacuum fitted, No.24 was commandeered to take the express on to Swindon. If operating to schedule, she would have been heading towards Bristol, and so would have had to haul the train tender-first.

In un-rebuilt form she was shown on GWR diagram 'N', in her rebuilt form on GWR diagram 'O'. The power class was 'B' and route availability uncoloured.

Ex-MSWJR 2-6-0 No.24 stood near the coaling stage at Swindon GW shed, c1928. ***F Moore.***

SHARP STEWART 4-4-4T Nos. 17 & 18

The MSWJR had aspirations to run fast through passenger services, so in 1897 they had purchased a pair of 4-4-4Ts Nos. 17 & 18 from Sharp Stewart. The locos won immediate acclaim for their appearance, but handsome is as handsome does, and performance-wise they were a dismal failure. Poor smokebox layout adversely affected their steaming, while having front and rear bogies meant that too little weight was carried by the coupled wheels, making them prone to slipping. With only 18 locomotives, there were times when the Company would have liked to have been able to use them for freight work because nothing else was readily available, but with brakes only on the coupled wheels they were dangerously prone to skidding if used on unfitted trains. Added to this, they suffered from overheated bearings. With all these shortcomings they were confined to secondary passenger services as soon as better locos became available.

In December 1923 No. 17 was working from Swindon Town, but moved to Andover at the end of the month. In March 1924 she was recorded as receiving her GWR number, 25. It returned to Swindon in May 1924, going to the GWR shed as the MSWJR shed had by then closed. In November 1924 she went into Swindon Works for repairs and although there for nearly a year, emerged with only minor changes; namely a new smokebox, GWR cone-shaped safety valve cover, raised bunker sides and GWR-style lamp irons. The extended smokebox would suggest that she was fitted with a superheater, but no record of this has been traced in the GWR records and no new diagram was issued. She was then put to work the Swindon Junction – Swindon Town shuttle, or the 'Old Town Bunk'. It is not recorded whether the new smokebox improved the steaming, but the locomotive still does not appear to have been a great success, as by October 1926 she had been stopped three times. So, when stopped for a fourth time in April 1927, it was laid aside and put on the *Sales List*. Not surprisingly, no-one showed any interest in buying a relatively large passenger tank, and she was scrapped in November 1928. The GWR diagram 'A' only showed her in pre-grouping condition, before the smokebox was extended and bunker sides raised in height. The power rating was ungrouped and route availability colour yellow.

No.18 was treated quite differently, as when the GWR assumed control of the MSWJR she was out of use awaiting overhaul, so was taken straight into Swindon Works. Emerging in February 1925, she had been given a GWR superheated taper boiler and a new cab, raised sides to her bunker, GWR-style lamp irons, and was renumbered No. 27. The new boiler would certainly have

Top: MSWJR 4-4-4T No. 17 at Cheltenham shed, c1923. *Photographer unknown.*

Middle: Ex-MSWJR 4-4-4T No. 25 at Swindon Works, around 1927. *H C Casserley.*

Bottom: Ex-MSWJR 4-4-4T No.27 following rebuild with GWR tapered boiler, c1927. *W Potter.*

improved the steaming, but sadly troubles persisted like her sister. After a brief visit to Cheltenham, she was transferred to Kidderminster, remaining there for the rest of her working life. The records show her stopped at Tyseley in August 1925 and sent on to Stafford Road Works for repair the following month. She was next stopped briefly at Worcester shed in February 1926, but then spent five months from October 1926 to February 1927 under repair in Worcester Works. Even so, she then needed further attention at Worcester, and was back there from May to October 1927. By the end of January 1928 she was back at Worcester shops yet again, a Kidderminster shed record states that she needed springs and crosshead spindle repairs. In June 1928 she failed at Salop with a hot box, and in September 1928 was back at Worcester for further repairs. Within two days of returning from the factory at the beginning of October, she had failed with a hot leading bogie box. This issue persisted and the locomotive ran a further hot box in February 1929. Finally in August 1929 she was sent to Worcester, and thence on to Swindon where she was condemned on 17 September. She was shown on GWR diagram 'B'. The power rating was 'A' and her route availability colour yellow.

BEYER PEACOCK 0-6-0 Nos. 19 - 28

In late 1899 the MSWJR purchased six mixed traffic 0-6-0s from Beyer Peacock. They proved very versatile machines, with a good turn of speed working passenger trains and adequate haulage power for freight. This led to four more being obtained in 1902. The one criticism levelled against them was the size of the boiler, which crews felt should have been larger.

After the war, the class appears to have been given priority for overhauls at Cirencester Works, and the entire class was in traffic at the time of the GWR takeover. The GWR allotted numbers 1003 - 1011 and 1013 to the class in MSWJR order, No. 1012 being unavailable as it was still carried by another GWR locomotive. By July 1925, No. 25 had been repainted in GWR green livery and given her new number 1008, but otherwise remained in MSWJR condition. A photo of No. 24 taken about the same time shows that she had been given a GWR parallel-sided copper-capped chimney, but otherwise remained unchanged. No. 28 remained in MSWJR condition until 1927 apart from the fitting of GWR numberplates.

Below: MSWJR 0-6-0 No. 22 at Andover Junction in 1921. ***A W Croughton.***

MSWJR 0-6-0 No. 24 in 1925 having been repainted in GWR green and given GWR numberplates as No.1008, but otherwise left unaltered. It is seen crossing Hatherley Road bridge at Cheltenham. ***H G W Household.***

After No.21 had been stopped at Cirencester in March 1924 it was sent to Swindon Works, emerging in January 1925 having been rebuilt with a superheated taper boiler and a new identity as GWR No. 1005. The rebuild must have been judged a success, as by December 1927 the entire class had been given similar boilers. Presumably the GWR agreed that the original boilers were on the small side. Apart from a brief visit by No. 1006 to Evesham in 1929 and spells at Kingham for Nos. 1004, 1005 and 1009 between 1932 and 1937, until 1936 the class generally remained allocated to their pre-grouping locations of Cheltenham, Swindon and Andover. They continued to be used on MSWJR line services, although they were also recorded as working on the Banbury & Cheltenham line and visiting Hereford and Leamington. In 1936 Nos. 1008, 1011 and 1013 were sent to Bristol St Philip's Marsh, where they remained until withdrawal.

Ex-MSWJR 0-6-0 No. 1005 shunting at Chipping Sodbury in 1925. ***H G W Household.***

The old order changeth, as an ex-MSWJR 0-6-0 shunts a GWR Siphon van at Marlborough around 1930. A few years before it would have been a parallel-boilered loco shunting MSWJR milk vans. ***Transport Treasury.***

The class remained intact until late in 1934, when Nos. 1004, 1006 & 1010 were all withdrawn. It became extinct with the withdrawal of the prototype rebuild, No. 1005 in the spring of 1938. The GWR 2251 class 0-6-0s, introduced in 1930 to replace elderly Dean and Armstrong locos, bore more than a passing resemblance to the rebuilt MSWJR locos, and it seems quite possible that they had some influence on the decision by the GWR to introduce the 2251 class.

The un-rebuilt locos were shown on GWR 0-6-0 diagram 'A19'. Oddly, the diagram showed the original rectangular cab sidesheets fitted to the first six members of the class, all of which had been removed well before the First World War. Their power classification was 'B' and their route availability colour yellow.

NORTH BRITISH LOCOMOTIVE COMPANY 4-4-0 Nos. 1 - 8 & 31

With the 4-4-4Ts failing to live up to their promise, the MSWJR had no locos really suitable for the fast through passenger services they had hoped to run. After careful consideration, in 1905 a single 4-4-0 was purchased from the North British Locomotive Company, which had been formed in 1903 by the amalgamation of Sharp Stewart & Co, Neilson Reid & Co (Hyde Park Works) and Dübs & Co (Queens Park Works). In the hope that the new loco would be the pride of the line, the first of the old SM&AR 0-6-0Ts was renumbered No.1A so that the 4-4-0 could become the new No.1.

This time the loco proved a great success, not only free running on the easy graded sections of line, but also powerful enough to take 300 tons up the 1 in 75 banks. Unlike the 4-4-4Ts it steamed well, showing good reliability with no reports of hot boxes. Over the next 7 years, six more similar locomotives were added to the fleet (Nos. 2, 3, and 5-8). A further pair (Nos. 4 & 31) were added in 1914, but differed in having a second dome on the boiler, incorporating a top feed fed by Gresham & Craven hot water injectors and surmounted by a pair of Ross 'pop' safety valves. The MSWJR was justifiably proud of their new acquisitions and always referred to them as the 'Number 1 Class'.

Unfortunately, 1905 was too early for the design to incorporate a superheater. The GWR did not begin trials of a Schmidt superheater until 1906, and it was 1909 before they produced the Swindon No.3 superheater. Although it was suggested that the later versions of the MSWJR 4-4-0s would be superheated, nothing was done. When the GWR assumed control in 1923, the fitting of superheaters to the class was seen as a highly desirable modification, and over the next five years the entire class received superheaters, some by rebuilding with GWR boilers, others by the fitting of superheaters into the MSWJR boilers.

The GWR allocated the class the numbers 1119 to 1126 and 1128 in MSWJR order, 1127 having been already allocated to the little Dubs 4-4-0 No.9. MSWJR Nos. 3 & 7 were due for overhaul and were taken in to Swindon Works in November 1923 and December 1923 respectively. When No.3 emerged from the Works in September 1924 as No.1121, their original boiler had been condemned and replaced with a GWR standard superheated taper boiler. On the other hand, when No. 7 returned to traffic in October 1924 as GWR No. 1125, she had been fitted with a superheater in her original boiler. She had also been given a new smokebox and a GWR chimney.

In March 1924 No. 31 was taken into Swindon Works and fitted with a GWR taper boiler and renumbered 1128. A superheater was then added to her original double-domed boiler, and in December 1924 MSWJR No. 1 appeared from Swindon Works as No. 1119 carrying this boiler in place of her original single-domed one. No. 1's original single-domed boiler was then overhauled and fitted with a superheater and put on to No. 8 when taken into the Works in March 1925, returning to traffic in October as No. 1126.

No. 8's boiler was the next to be overhauled and superheater-equipped, before fitting to No. 5, after it had been taken into Swindon and her original boiler condemned in September 1925. No. 5 returned to service in February 1926 as No. 1123. However, after being stopped in May 1929, it was then rebuilt with a GWR taper boiler before returning to traffic in the September. Her old MSWJR boiler was given a new firebox backplate and new superheater tubes, and in due course fitted to No. 1119 to replace the double-domed boiler when she went to Swindon for general repair in the winter of 1931/ 32.

No. 6 was renumbered 1124 in March 1924, but otherwise appears to have remained unchanged until Swindon supplied new cylinders, which were fitted at Cirencester Works in September 1924. Even then she probably remained little altered until July 1925, when upon arrival at Swindon the MSWJR boiler was condemned and replaced with a GWR tapered version. No. 1124 re-entered service in January 1926.

Except for the fitting of GWR No. 1122 numberplates, MSWJR double-domed loco No. 4 remained in original condition until May 1927, when a superheater was fitted to her original boiler. Single-domed No. 2 was repaired and fitted with new cylinders at Cirencester Works in May 1924 and renumbered 1120, after which she remained virtually unaltered until taken into Swindon Works in August 1927. Her original boiler was then condemned, and replaced with a GWR superheated taper boiler, returning to service in March 1928.

The GWR had now achieved its target of superheating the entire class, four with GWR boilers and five with modified parallel boilers. But as this had taken over 4 years, it was not long before the earliest conversions became due for overhaul. Parallel boilered No. 1126 was stopped in July 1928, rebuilt with a GWR taper boiler and re-entered service in November 1928. The rebuilt No. 1121 had her taper boiler replaced in March 1929, after being stopped in October 1928. In 1930, No. 1122 had its double-domed MSWJR boiler replaced with the single domed boiler which had originally been fitted to MSWJR No. 1, but which had been transferred to No. 1126 after superheating. Both parallel boilered No. 1123 and taper boilered No. 1128 went in to Swindon Works for heavy repair in May 1929, both emerging in September 1929. No. 1123 was released with a taper boiler in place of her old MSWJR parallel one, whilw No. 1128 just exchanging her GWR taper boiler for a different one. This meant that there were now six locomotives with GWR boilers, leaving just three with parallel units.

As No. 1120 was not rebuilt until 1927 it was not due for overhaul when she was stopped in June 1931, but apparent underframe problems caused her to be condemned in the August. The un-rebuilt No. 1125 was stopped at Gloucester in August 1931, and was eventually condemned at Swindon in February 1932. No. 1123's old boiler was repaired and eventually fitted to No. 1119 when it was overhauled in the winter of 1931/32. No. 1124 did not receive a boiler exchange and general repair until the summer of 1932. There were then no further withdrawals until 1935, when the two surviving parallel boiler locos, Nos. 1119 & 1122, and taper boilered locos Nos. 1124 & 1128, were all condemned and scrapped. No. 1121 went in the following March, No. 1123 in April 1938, and the class finally became extinct when No. 1126 was condemned in December 1938.

MSWJR 4-4-0 No. 2. ***Author's collection.***

Apart from brief visits to Gloucester, the class remained allocated at their pre-grouping locations of Cheltenham, Swindon, and Andover (although the Swindon shed was now the GWR one). They all continued to be used on MSWJR line services, although some were recorded as visiting Hereford and Leamington. The unmodified locomotives were shown on GWR 4-4-0 diagram 'A35', the superheated examples which still had MSWJR parallel boilers on 'A38' and the engines with GWR taper boilers on 'A37. Their power classification was 'A' and their route availability colour yellow.

Above: MSWJR 4-4-0 No. 31, with parallel boiler and fitted with a second dome. *Author's collection.*

Below: Ex-MSWJR 4-4-0 No. 1119 at Swindon Works after being given the double-domed boiler off MSWJR No. 31. *Author's collection.*

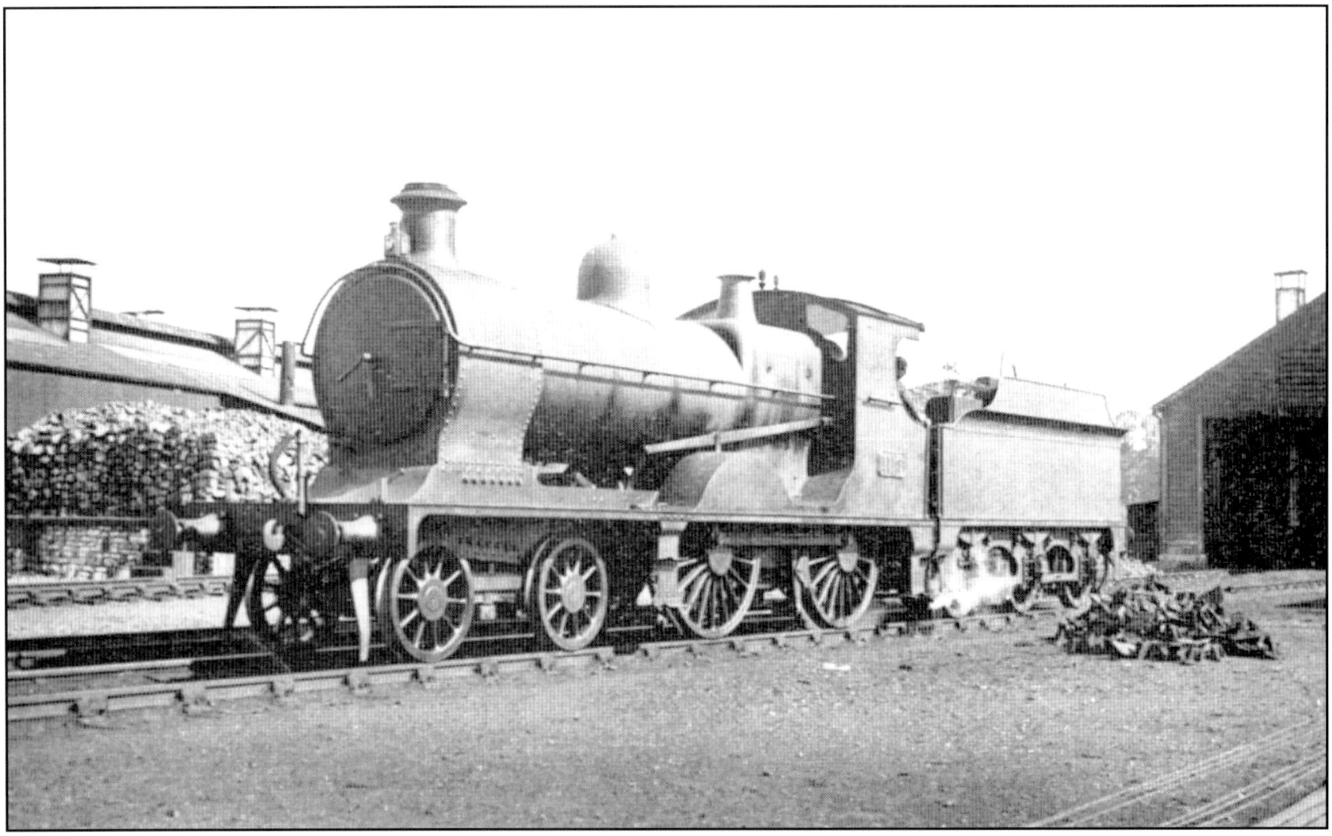

Above: GWR No. 1123 (ex MSWJR No. 5) stood at Andover Junction shed c1928, still with a parallel boiler but with Great Western style safety valve bonnet fitted.

Below: Ex-MSWJR 4-4-0 No. 1124 at Southampton after rebuilding with a GWR taper boiler. Both images *Author's collection.*

BOOK REVIEW

A MALMESBURY RAILWAY COMPANION:

A COMPENDIUM OF STORIES AND DELIGHTS RELATING TO THE MALMESBURY BRANCH RAILWAY

MIKE FENTON

Lightmoor Press (ISBN 978 191506 9146)
Casebound, 200 pages, 273mm x 215mm, Design by Adrian Knowles
£35.00

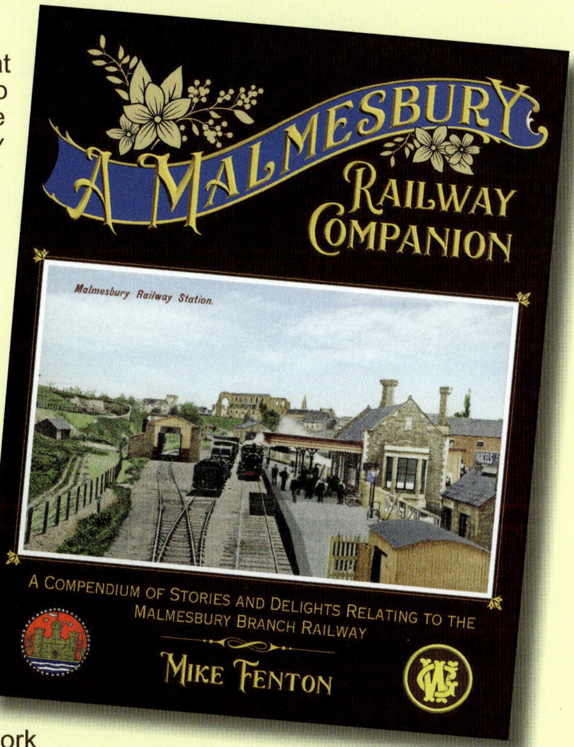

The name Mike Fenton should be familiar to followers of Great Western branch lines. The name Wild Swan will also be familiar to the same audience. The two came together in 1990 to produce one of the many iconic Wild Swan branch line histories, *The Malmesbury Branch*. Three and bit decades later, substitute Wild Swan for Lightmoor Press and we have an additional work by the same author on the same line and in the opinion of your reviewer, every bit it as interesting.

What this new book does is place the railway into the community through the eyes and memories of those who used it, saw it, and worked on it. It is a book full of unashamed nostalgia yet I also read it with a growing sense of sadness. Sadness as this is an account of a way of life not just fast disappearing but no, all but disappeared. Before I reached the end I had also come to the same conclusion as Mr Fenton below.

Towards the end of the work the author makes the point, 'We are also, I feel, at the end in another context. It is unlikely that there will be many more books on old branch railway lines…'. He goes on to explain this rationale as so few staff now remain who will have worked on and recall the lines that closed in the 1950s and 1960s. True the bold facts, dates / figures etc will survive, but it is the human element that has gone; what made it all 'tick'.

In that respect, the lines of the former GWR and SR have undoubtedly fared better than most, whilst even if a new in-depth work were published on say an obscure LMS / LNER line today the question has to be asked would it even be considered commercially viable? The railway author can spend years in time, a considerable amount of finance in travel, photographs, etc and for what return. Certainly not a financial one.

If all this sounds depressive it is not meant to be. Here is a book full of unashamed nostalgia and Messrs Lightmoor Press are to be congratulated in rising to the challenge of ensuring the detailed research it contains is now available to a wider audience. It would be interesting to speculate how well it might sell in Malmesbury itself, where I suspect most of the current residents will never have known the railway.

Mr Fenton also makes the point there are still gaps in his studies, some of these photographic. He again raises a question what to do – and how to bring these to a wider audience – if they were to appear. He doubts there will be a third book, but may we respectfully suggest *Western Times* as a potential alternative output medium if one is needed. I would simply say, well done Mr Fenton, well done Lightmoor Press – and please do not stop looking.

Returning finally to the book, although predominantly monochrome there is a short section in colour at the rear. A pity perhaps there was similarly not odd colour pages elsewhere, such as on p.35 where the labels concerned would have been better in their original colours. A minor criticism perhaps and it can be done by having odd colour sections within an otherwise black and white book.

Overall then a well designed and I would say reflective book, full of local nostalgia that left your reviewer hankering for the past – whilst still typing this on 21st century technology. *KJR.*

Top: On a rather murky 4 March 1961, Collett 0-4-2T No. 1421 has arrived at Marlow on an early morning train, with the auto-coach unusually marshalled at the rear of the freight stock. Consequently, this was probably an Empty Coaching Stock movement so far as the passenger vehicle was concerned. The goods wagons will be shunted into the yard before the locomotive is reconnected to the trailer for the passenger service to commence. The lack of signalling infrastructure in the station environs is clearly evident. *James Harrold (H2013).*

Middle: Having disposed of its train, it is now time for the engine to receive attention with both crew actively engaged. The amount of physical labour involved is often forgotten. Coal would first be shovelled from a wagon onto the coal staithe, and later into the bunker. At a later stage, the process would be reversed with ashes cleaned from the locomotive shovelled from ground level back into the now empty wagon for removal. Considering the effort required, the change to diesel traction was understandably welcome for many.

The branch's ability to accept large engines on excursion workings is confirmed by the wording on the notice board beside the shed entrance: 'Engines in the Red and Blue classification must not pass these doors.' Presumably these excursions related to river cruises as Marlow would have been a convenient starting point for such adventures. *James Harrold (H2015).*

Bottom: Later that day, No. 1421 awaits departure with the 10.05 am service to Bourne End and in more conventional mixed-train formation, empty wagons are marshalled at the rear. Although out of shot, a Toad would be necessary at the rear of the train.

Permanent way work is in hand judging by the sleepers stacked on the platform. The generously-sized goods shed is in the left background with a light truck (a Morris Commercial?) in attendance. The adult on the platform seems to be conversing with the locomotive crew while holding the hand of a youngster. How often did such encounters first ignite the flame of youthful enthusiasm? *James Harrold (H2016).*

THE MARLOW BRANCH

ONE THAT MANAGED TO GET AWAY

One hundred and fifty years ago, the 2¾-mile branch from Bourne End on the Wycombe Railway (Maidenhead to Thame via High Wycombe) to Great Marlow was opened. No doubt in 1873 there were many toasts to the new venture's continued success, but almost 90 years later it was under serious threat of closure as a consequence of the Beeching Report.

Fortunately Transport Minister Barbara Castle refused to sanction closure of the short branch to Marlow (as it was renamed in 1899) although other parts of the original Wycombe Railway's route (Bourne End - High Wycombe and Princes Risborough - Thame) were less fortunate. The Bourne End to High Wycombe section alone was reported in November 1968 as losing £60,000 annually. Closure was unsuccessfully challenged, not least in the interest of eighty schoolchildren who used the service daily as there was no alternative bus service. Closure notices were posted the following month.

One legacy of closure north of Bourne End was that Marlow-bound services could only arrive from Maidenhead and as the branch trailed in from the south, a change of direction became necessary. A similar situation exists on the truncated remains of the Calstock branch from Bere Alston on the ex-Southern Railway network.

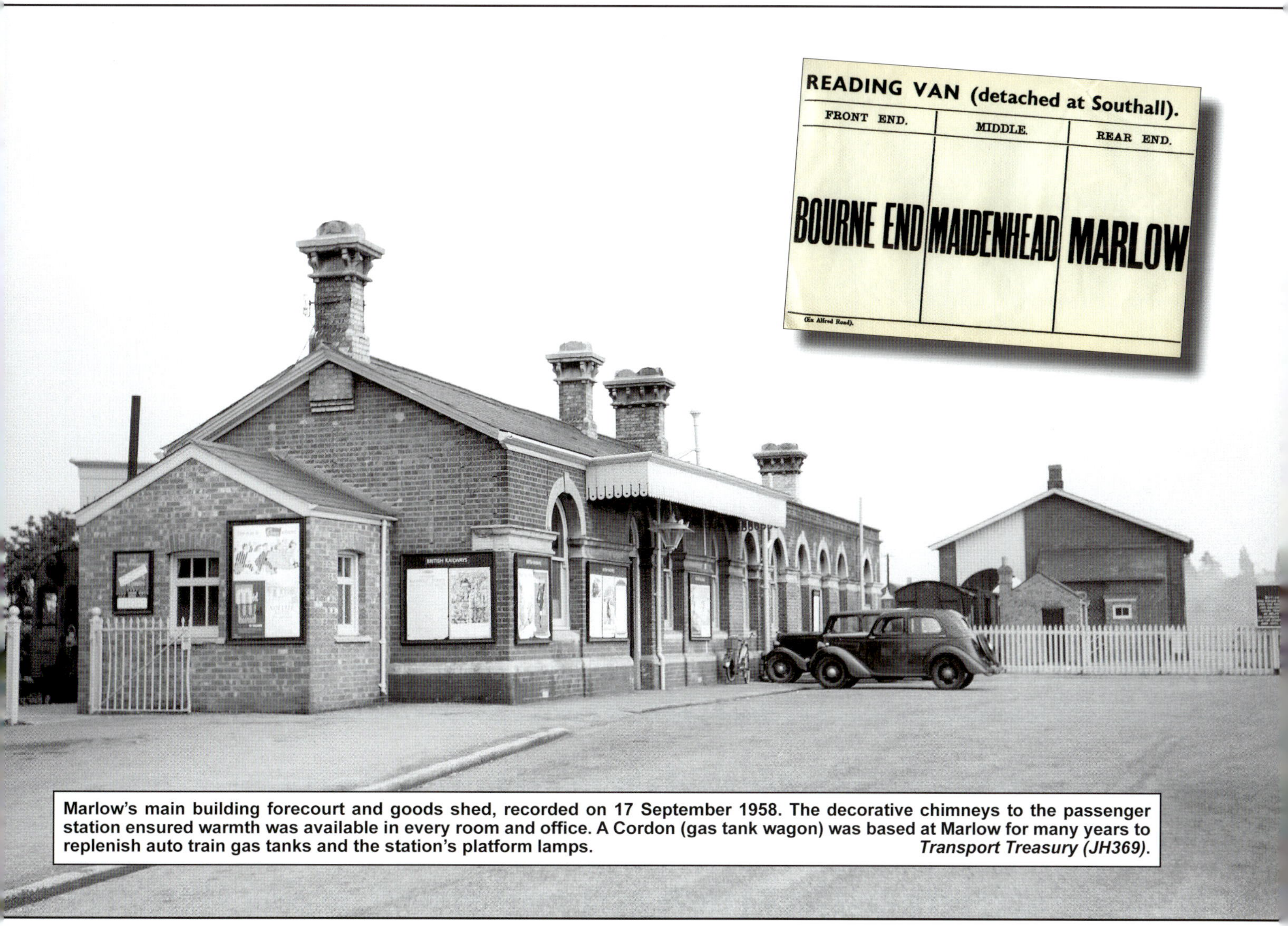

READING VAN (detached at Southall).

FRONT END.	MIDDLE.	REAR END.
BOURNE END	MAIDENHEAD	MARLOW

(Ex Alfred Road).

Marlow's main building forecourt and goods shed, recorded on 17 September 1958. The decorative chimneys to the passenger station ensured warmth was available in every room and office. A Cordon (gas tank wagon) was based at Marlow for many years to replenish auto train gas tanks and the station's platform lamps. ***Transport Treasury (JH369).***

Above: The large timber yard adjacent to the station with a French-built steam crane in use. This area became the site of the new single platform station opened in February 1967. *Transport Treasury (H372).*

Below: The physical end of any hoped-for extension towards Henley. This view dated September 1958 shows that the cattle pens have had little recent use. The milepost to the right records 2¾ miles from Bourne End. GWR spear fencing in BR cream livery protects railway property and the bullhead rail is supported by standard GWR 2-bolt chairs. *Transport Treasury (JH370).*

Around the time that Marlow branch closure was mooted, Eric Miles began taking an interest in the station and the local railways. He started to gather snippets of information which might otherwise have been lost and in consequence recorded folklore from the 'old railway's' final days. In that context, 'old' meant the branch prior to its slimming down to the present day basic railway. His interest also broadened into assimilation of information about Bourne End, and how local services integrated into operation of the general network. Among the stories that Eric Miles gathered:

- In the latter part of the 19th century, proposals were formulated for construction of a railway line to connect the branch termini at Marlow and Henley-on-Thames, over a distance of about 8 miles (by road) on the north bank of the river. The opposition of the Leander Rowing Club at Henley would sink these plans. (This is one of the world's oldest rowing clubs and presumably wielded considerable influence).

- In the spring of 1947, the local name for the branch service, the 'Marlow Donkey', received national coverage when it was broadcast on the BBC radio programme *In Town Tonight*, then a popular Saturday evening radio series. At that time, Bourne End-Marlow services were being operated by two 'Clifton Downs' auto coaches. During the broadcast, BBC reporter Tony Rich interviewed Miss Cecily Bateman whose job it was to keep these vehicles clean.

Miss Bateman told listeners that she started work at 7.00 am to prepare the train for its first trip at 7.30 am. She subsequently made a number of trips between the two stations, '.... taking every opportunity to wash and dust the compartments, to polish door handles and other brass work, and to keep the windows clean'. The result was that a trip on the Marlow Donkey was a real joy for the passengers. Miss Bateman went off duty at 3.00 pm, stating she 'enjoyed her job which she had held for 7 years'.

Left: An undated view from the late 1950s shows Collett 0-4-2T No. 1445 heading into the bay platform at Bourne End with a train from Marlow. The connection to the main line is via the left hand track. The curvature is sharp enough to warrant a check rail on both lines plus a 5 mph speed limit (the notice on the post is indecipherable even under magnification). In the junction 'V' is Bourne End yard which hosts a corridor coach and an auto trailer, ready to handle a sudden increase in demand. Two mineral wagons and a van await collection in the right-hand siding. There is much detail to note in this view looking south towards Cookham and Maidenhead: round point rodding; a gas lit lamp in the foreground; wooden walkway over the tracks; two adjacent trap-points to protect the running lines against runaways from the sidings. ***Transport Treasury (LS716).***

Right: One of the Hawksworth-era auto trailers with motive power provided by No. 1448 (unseen at the far end) is standing in Bourne End's curved bay platform. The considerable gap between the platform and centre door is the reason why passengers had to use the driving cab or guard's compartment doors for access. The problem was less acute with the previously allocated 'Clifton Downs' stock which was 50' or 51' long. The BR-built auto trailers were 63' in length. ***Transport Treasury (H374).***

Work-stained 2-6-2T No. 5190 is involved in shunting the yard at Bourne End on 17 September 1958. Such operations could cause considerable delay to road traffic although today on this side there is just a Ford Consul waiting (note the splendid late series Austin 7 saloon parked across the road). This view is looking north towards High Wycombe with the vintage wooden bracket signal affording departure from the right-hand platform for 'wrong line working' on to the single line while the smaller arm controlled access to the goods yard. If the engine is intending to reverse into the yard, by the regulations it should move forward and wait this side of the signal. However, unofficial practice at many locations meant that the signalman would change the road and lower the signal once the yard turnout had been cleared even if the locomotive was not fully past the post. ***Transport Treasury (H375).***

- The *Model Railway Constructor* for July 1954 included an update on the Clifton Down stock. 'For years the service was operated by pair of auto-fitted Clifton Downs coaches. Alas, the local St Trinian's did so much damage that these stalwarts were withdrawn and one of the only too common trailer cars used instead. The reason for the use of the Clifton Down coaches, had been the excessive curvature of the bay line at Bourne End. Only the end doors can now be used on the standard trailer.....' (The 1954 article may have been a little behind the times as it is likely the Clifton Down stock had ceased to run about 1950).

- For a time prior to 1936 a steam railmotor was also used, while before any form of auto working, a gravity shunt was necessary to transfer the engine to the Marlow end of the train after arrival at Bourne End. 'My grandfather remembered the occasion when the coaches came running back into the station, the engine backed on and after getting the right away, left for Marlow .... by itself as no one had coupled up. I believe this was the incident which caused the name 'The Moke' instead of the 'Donkey' to be used'.

Editors note: According to *The Oxford Modern English Dictionary*, a 'moke' is a donkey or a very poor horse.

- In 1954, new signalling at Bourne End enabled bi-directional working north out of the station and also on to the branch. Marlow signal box was closed that year, signals were removed, and the branch thereafter operated on the 'one engine in steam' principal. A wooden staff was provided on which was a key for unlocking the ground frame at Marlow for run-round and shunting purposes. Aside from mixed and goods workings, the normal passenger service was covered by auto trains. At Bourne End the branch train used the main platform whenever possible to avoid difficulties with the curved bay.

- In the early 1960s and prior to steam replacement the number of trains approximately equated those of ten years earlier. These included a daily freight, Monday to Friday, south from High Wycombe which would shunt coal at Woburn Green; empty wagons were then set down at Bourne End before continuing to Maidenhead and finally Slough.

- Another goods working brought coal from Slough, setting down at Bourne End and then running around before continuing to Marlow. A substantial timber yard existed at the terminus which received round timber with heavier sections on bolster wagons. The timber yard was probably one of the reasons for cancellation of closure plans. Scrap metal from the Ford plant at Langley was also routed to Marlow but freight traffic ceased completely from 7 July 1966. Goods traffic to Bourne End ended on 11 September 1967 while the yard at Cookham on the line south to Maidenhead had closed in 1965.

- After shunting at Marlow, the train would return to Bourne End where the coal yard was shunted and empty wagons from Marlow left in the yard. Then the train went to Woburn Green 1½ miles up the Wycombe line to collect empty wagons before returning to Slough, picking up the previously deposited empties at Bourne End.

- Marlow branch steam motive power was invariably a Class 14xx with the service occasionally extended to Maidenhead, dependent upon the day and the season. The first and last trips between Bourne End and Maidenhead could also run as ECS workings. Steam on the Maidenhead-High Wycombe line was usually in the form of Class 57xx 0-6-0PTs – later replaced by 94xx series – and Class 61xx prairies. Freight workings brought a variety of engines including at least one Class 55xx 2-6-2T on a Marlow turn in 1964.

- Rationalisation really began with the 1954 signalling changes. After a hiatus of five years, the next change was introduction of second class only on the branch. Steam was replaced on the 'Marlow Donkey' service from 8 July 1962 with a single car diesel railcar. This was considered adequate for off-peak hours while a spare driving trailer was held available in a siding at Maidenhead rather than Bourne End, which would have been more logical. Marlow was still a staffed station but with effect from 31 January 1966, staff were only present up to 0900 hrs. After that the Guard issued tickets on the train itself.

- From 31 January 1964 a through Slough to Crewe fitted freight, referred to as the 'Chilton Night Trader' although it is unclear whether this was an official or local title, ran northward over the Wycombe railway. Castle Class No. 7013 appeared on this train once, devoid of its *Bristol Castle* nameplates. Other unusual visitors included Eastern Region classes B1 and V2 which travelled south certainly as far as Bourne End on passenger excursions, usually referred to as 'River Cruises'. County Class No. 1011 *County of Chester* was also seen on 28 February 1962 with a special from Bristol to Bourne End. These cruises would start from Marlow with the main line train collecting its passengers again either at Henley or, as in the case with No. 1011, at Windsor.

At a later date to the previous image, the wooden bracket has been replaced by a standard tubular metal post with an 'elevated' disc for the goods yard in place of the former arm. This was standard practice on the Western Region in the mid-to-late 1950s as signals became due for renewal. This photograph shows so much of the operational 'furniture' that was common in the 1950s: the rodding leading to the gate bolt in the six-foot; standard sign requiring passengers to cross the line by the footbridge; token exchange apparatus/ bull's horn in front of the signal box; spear fencing in cream; trap points providing flank protection for the yard exit; plethora of signals in the distance; and a Morris Minor crossing the railway. ***Transport Treasury (1971).***

Above: Class 94xx No 9415 is arriving wrong road at Bourne End on a freight working at 11.42 am on 4 March 1961 while one passenger watches from the other platform. The round timber on the bolster wagons is probably destined for Marlow. Another everyday scene now consigned to history which displays the GWR-design covered footbridge, the bracket signal behind, and the platform barrow and trolley. *Transport Treasury (H2022).*

Below: Loudwater on 4 March 1961 with Auto trailer Diagram A38 No. 229 in the foreground. Ignore the coach's BR ancestry plus the seat support at extreme right and this could be a pure GWR view with lamps, fencing, pagoda shelter, station signage, platform trolley, signal in the background – and the seat on the other platform. *Transport Treasury (H2030).*

Above: The end of the line at High Wycombe, with a service from Maidenhead or Marlow in the bay on 3 March 1957. Most off-peak services were well served with just a single coach although there was also at least one mid-day through working from Paddington. No. 1448 has the tail lamp in the correct position and so is ready for departure. The auto trailer is definitely a 70' Diagram L and might be the same vehicle as that in the hands of No. 1450 in the photograph above at Bourne End. ***Transport Treasury (LN671).***

- A bizarre episode was reported in *The Railway Magazine* for January 1962 when 'Railway Hotel' in Marlow changed its name to the 'Marlow Donkey'. The occasion was apparently seen as an excuse for 'general jollification' including a number of people in fancy dress. The full story subsequently emerged in Court proceedings. Three men who admitted 'having had an amount of drink' had walked half a mile up the branch, set up three detonators, and held up the 8.10 pm 'Donkey'. On stopping, the driver was confronted by the men, two of whom were masked, wearing cowboy outfits and complete with six-shooters. Neither the driver nor British Transport Police saw anything amusing and the culprits were brought before the Court where they pleaded guilty to 'Entering a carriage other than on the side of a platform', 'Unlawfully and wilfully stopping a vehicle on the railway,' and 'Wilfully obstructing the driver in the execution of his duty' all of which were offences under railway by-laws. It was elicited that the train had only one legitimate passenger. Two of the accused were members of a jazz band who had intended to meet the train in fancy dress. The Court was also not amused and each 'cowboy' was fined 30/-. After leaving court they reputedly adjourned to the Railway Hotel/ Marlow Donkey for a celebratory drink!

- Back on the ground, diesel power also replaced steam for the final years between Bourne End and High Wycombe with a four-car train necessary during peak periods. A simplified station layout was provided at Loudwater with a single platform and reduced siding accommodation. With cessation of freight services from 10 February 1967, Marlow was reduced to a single platform located on the site of the erstwhile yard and the former GWR station structures were then demolished.

- Steam briefly returned to Marlow in the summer of 1973 when 14xx tank No. 1450 was brought from the Dart Valley Railway and sandwiched between two auto trailers for a Great Western Society special to commemorate the branch's centenary.

The present service pattern is Maidenhead to Bourne End with reversal and continuation to the austere and basic terminus at Marlow. Current train formations lack the interest of past years but at least a service survives. Is it too much to hope that in these enlightened times, a link north from Bourne End might be restored?

EXPERIMENTAL MOTIVE POWER:
BROAD GAUGE *HURRICANE* & *THUNDERER*

In a career of unprecedented pioneering achievement, Brunel's dictates on locomotive design are regarded as one of his few failures. The difficulties experienced in operating services during the Great Western's earliest days have been held as proof of shortcomings as a mechanical engineer which seems unreasonable considering his three ships which were enormous advances in naval architecture and construction.

Despite the ostensible impracticalities of his steam locomotive specifications, four of the contractors invited to tender responded by preparing designs, followed by actual construction and delivery to the fledgling company. With the railway revolution gathering momentum, it is hard to imagine that they were short of work but the Great Western project was a grandiose enterprise with which they were presumably keen to be associated. The broad gauge was dimensionally so different as to raise unprecedented issues beyond the borders of the slim empirical data gathered thus far from standard gauge practice. On the other hand, this was a period when by definition many locomotives were experimental in nature and the spirit of innovation in the minds of talented and creative individuals would have encouraged some to attempt what with hindsight was to be deemed impossible.

With regard to the fifth contractor, there is circumstantial evidence to suggest that Brunel might have been more intimately involved in the supply process beyond issue of invitations to tender.

Contract Specification

In planning the initial locomotive fleet, the specifications issued to prospective tenderers were challenging. The opening sentence referred to the time needed to supply one or two locomotives. The following sentences afforded latitude to the contractors for development of their own ideas for technical advance while also providing the basis for series production. Thus far, the terms were eminently sensible. With an unusual gauge and with a brand-new trunk railway line in creation that sought high standards, to canvass widely would hopefully focus the best technical minds on optimised solutions. The notion of more orders deriving from the most practicable designs would have been a powerful incentive. With so much yet to be learned, it is hard to think of a better means of concentrating the best minds on the case.

The third paragraph started reasonably by nominating 30 mph as the 'standard or minimum velocity' but then raised the first hurdle in limiting the rate of piston travel to a maximum of 280 ft per minute at this speed. The next paragraph stipulated a maximum boiler pressure of 50 lbs/ sq in and a force of attraction (presumably equivalent of tractive effort) of 800 lbs on level track at 30 mph. There then followed a weight limitation of 10 tons 10 cwt for the locomotive in working order excluding the tender, with a stipulation of six wheels if the weight exceeded 8 tons. The final dimensional requirements stated the gauge at seven feet, with 'the height of the chimney as usual'.

The weight restrictions appear to have been attempted insurance against the fragility then obtaining in construction materials. This was around 18 years before patenting of the Bessemer process which would set new boundaries in steel quality and reduced production cost. A slow piston speed seems to have been stipulated as a means of reducing risk of metal fracture. Early broad gauge designs embraced significant safety margins and perhaps laid the foundation for the safety measures that formed such an important component of the company's lasting culture.

These terms presented conflicting and ultimately insurmountable challenges. The nominal operating speed was acceptable in contemporary terms but the means of its achievement while working under load

This fine model of *Hurricane* is on display in the museum of the Great Western Trust at Didcot, and was built by the late Mr Harper who passed away about 40 years ago. ***Peter Rance, GWT.***

were virtually impossible. The slow piston rate required a short stroke and/ or large diameter driving wheels. At the time, other engineers were exploring the possibility of short piston strokes but the concept was soon abandoned. Interest in the idea presumably stemmed from the desire to reduce stress on components whose materials might contain casting imperfections.

An engine operating within the maximum permitted piston rate would require large diameter, and thus heavy, driving wheels which would demand sacrifice of weight and strength in other parts of the design. It is hard to accept that a man of Brunel's abilities had overlooked this mutual incompatibility but subsequent correspondence suggests that he resorted to a bombastic tone with contractors, consequent upon their inability to deliver satisfactory locomotives. His apparent attempt at denial of responsibility seems hardly to have been to his credit.

Paddle Steamer Influence?

Broader issues may have prevailed upon Brunel's thinking. He was concurrently engaged in creating a modern, integrated transport link between London and New York, the scale of which must have been breath-taking at the time. This programme contemplated operation of a trans-Atlantic shipping service using technology erroneously believed impossible by many. Against the magnitude and complexity of those objectives, moving trains over the 100-odd miles between London and Bristol could well have been dismissed as a trivial element in the total scheme.

The new service contemplated speedy conveyance of first class passengers in great comfort, as apparent in the design of the *Great Western*, a wooden-hulled paddle wheel steamship equipped with masts and sails to cover the transatlantic leg. There was widespread scepticism surrounding the project's feasibility, deriving from the belief that the vessel would be too large and therefore unable to carry sufficient fuel to make the crossing under steam power alone. However, Brunel understood a fundamental point that others had failed to appreciate i.e. that the ship's capacity increased as the cube of, whereas the water resistance against the hull increased as the square of, its cross-sectional dimensions. Thus, a larger ship was naturally more fuel-efficient, as proven when *Great Western* completed its inaugural Bristol-New York crossing in April 1838 with fuel to spare. The dates are significant as the vessel was laid down on 26 June 1836, launched on 19 July 1837, and completed on 31 March 1838 i.e. within the time frame of design and construction of the initial locomotive fleet.

Other factors were relevant to the ship's comfort and efficiency. The masts and sails enhanced speed and fuel consumption under favourable weather conditions while in rough weather also helped to maintain an even keel and thus stability. Reduction in the propensity to roll reduced the risk of paddle wheels rising clear of the waterline. A paddle wheel intermittently free of water resistance generated unequal stresses on the power transmission, while also inducing the ship to change course. Constant immersion of both paddle wheels was thus desirable to maintain steady forward momentum.

Another feature that contributed to stability was the use of a pair of side-lever engines constructed by Maudslay, Sons & Field [with whom Charles Collett would commence his career many years later]. This type, common to most early steam-powered sea-going vessels, was a development of the beam engines used by early pioneers such as Newcomen and Watt to drive mining pumps. These were large with the particular advantage of having a low centre of gravity. The revolutions per minute of the paddle wheels were substantially fewer than would later be the case with the propellers of screw steamers, and thus ideally suited to the slow piston speed of side-lever engines.

Right: The *SS Great Western* was Brunel's first steamship, laid down 26 June 1836, launched 19 June 1837, and completed fitting out 8 April 1838. Built by William Patterson of Bristol, she initially provided a passenger service between that port and New York. With displacement of 2300 tons, gross registered tonnage of 1340, length at 235 feet and beam across the paddle wheels of 57 ft 9 in, she was a considerable quantum leap in passenger vessel size by the standards of those days.

WESTERN TIMES

If overcoming the resistance towards his novel ideas for *PS Great Western* was uppermost in his mind, Brunel might have simply translated a marine-derived specification to his proposed railway locomotives, without thought for the consequences. Certainly, the concept of a low centre of gravity seems to have figured large as evidenced by his early ideas for passenger carriages, which were to have four feet diameter wheels with the body positioned between them. Perhaps, this was also the reasoning behind the stipulation in the contractors' specifications of a gauge of seven feet and "the height of the chimney as usual", a combination which might have implied a low centre of gravity. In this respect Thomas Crampton, a near contemporary, was also much exercised by this factor, as apparent with the locomotive type that he developed. This ultimately proved largely unsuccessful in the UK but gained quite wide acceptance in continental Europe.

Pursuing the themes of contemporary marine practice and the unprecedented demands of a trans-Atlantic transport system, there was another factor that might have influenced Brunel's thinking. There are less constraints in the design of the power unit for a steamship as the engine room volume is essentially cuboid, and restricted only by the overall dimensions of the hull – principally the beam. In contrast, the corresponding volume with a steam locomotive is awkwardly shaped being long and narrow, and hemmed in by track and loading gauge dimensions. This hypothesis is speculative and based on the sequence of events during the period that the initial locomotive fleet was being planned. The specifics of how and why Brunel claused the contractors' invitations in this form remain unclear but it seems possible that facing so many other pressing issues, he considered this subject to be of secondary importance. On the other hand, the challenges facing the contractors were acute. The resultant machines were definitely experimental but may not have seemed so outlandish then as they do now among the novel attempts of contemporary engineers to gain commercial advantage.

THE LOCOMOTIVES

Nineteen were provided to meet the company's initial requirements:

Builder	Location	Introduced	Makers Nos.	Names
Charles Tayleur & Co	Newton-le-Willows	1837 1838	51-3 62-4	*Vulcan, Aeolus, Bacchus* *Appolo, Neptune, Venus*
Marther, Dixon & Co	Liverpool	1837/ 8 1839/ 40	40/1, 50, 53[?], 51/2	*Premier, Ariel, Ajax* *Mars, Planet, Mercury*
Sharp, Roberts & Co	Atlas Works, Manchester	1838	R, S, T	*Lion, Atlas, Eagle*
Haigh Foundry	Wigan	1838	25/6	*Snake, Viper*
R & W Hawthorn & Co	Newcastle	1838	235/6	*Thunderer, Hurricane*

ET MacDermot stated 'in the whole history of British railways there has never existed such an extraordinary collection of freak locomotives as those that were built ...and delivered during a period of about eighteen months from November 1837'. Except for the products of R & W Hawthorn & Co, all were 2-2-2s with driving wheel diameters varying between 6' and 10'. The first four builders on the list evidently operated in an open tender situation but the circumstances surrounding the Hawthorn-built pair seem to have been different.

Thunderer and *Hurricane* were built under a patent taken out by one TE Harrison in December 1836. The essential feature of the idea was that the boiler should be carried on a 6-wheeled vehicle (the boiler unit) separated from another (the engine unit) which accommodated the cylinder block, valve gear, and driving axle[s]. The cylinders were fed by a horizontal steam pipe which passed through the smokebox front and exhausted by way of a return pipe for discharge through the chimney. The live and exhaust steam pipes had flexible couplings to accommodate the articulated connection between driving and boiler units but information on the nature of these joints is lacking. Considering the difficulties experienced later with pressure seals in the 'Atmospheric Caper' it may be assumed that these couplings were a source of leakage. The tender was typical of the genre at that time and was positioned at the footplate end of the ensemble. The fireman rode on the footplate while the driver rode on the driving unit so as *inter alia* to adjust the valve settings as required.

The tenders were identical, as were the boiler units except that *Thunderer*'s chimney was 16" in diameter while that for *Hurricane* was 2" greater. The engine unit for *Thunderer* had an 0-4-0 wheel arrangement with spur gears between connecting rods and driving axle, details of which have apparently not survived. However, the 27:10 gear ratio increased the actual driving wheel diameter of 6' 0" to the equivalent of 16' 3". [The idea of spur gears was more or less contemporaneously disproved with engines *Snake* and *Viper*]. However, *Hurricane*'s driving unit was a 2-2-2 with conventional transmission using connecting rods. The similarities and differences were evident through the line drawings published in *The Locomotive* for April 1901, Volume 6 No. 64.

Broad gauge locomotive *Hurricane* as delivered to the GWR by Hawthorn on 6 October 1838 with what might be considered a 2-2-2+2-2-2 wheel arrangement. ***The Locomotive.***

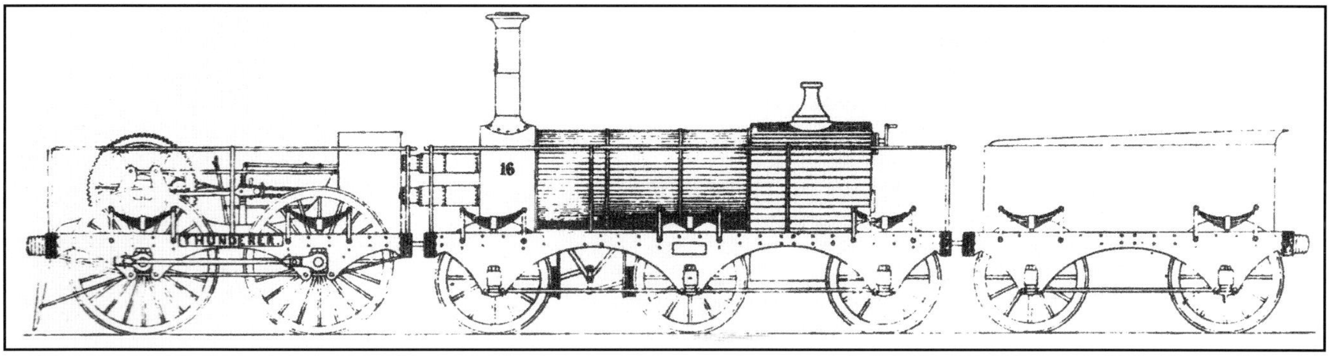

Thunderer* was delivered by the same manufacturers on 6 March 1838. The pair were similar in most respects except for the wheel arrangement and the nature of the power transmission between cylinders and driving wheels. The exact configuration of *Thunderer*'s spur gear arrangement apparently remains indeterminate.** ***The Locomotive.

The background to these machines apparently differed through the nature of the relationship between Brunel and Thomas Elliott Harrison [1808-88]. Following a grammar school education, Harrison was apprenticed to civil engineer William Chapman and worked initially on dock construction. He then assisted Robert Stephenson on plans for the London & Birmingham Railway, and later became engineer on the Stanhope & Tyne Railway (a major operator of inclined planes) followed by an appointment as engineer [1834-8] of the Durham Junction Railway (which included construction of the Victoria Bridge over the River Wear). He worked again for Stephenson on the Newcastle-Berwick route, and then became Chief Engineer for the York, Newcastle & Berwick Railway. The company became a major constituent of the North Eastern Railway where he served as Chief Engineer from 1854 until his death.

Despite this experience which reflects a career on the civil side, Harrison was acknowledged as an engineering all-rounder and his time on the Stanhope & Tyne Railway seems to have induced lateral consideration about alternative means of optimising motive power configurations. In any event, according to *The Locomotive Magazine*, he had previously designed locomotives to the general principles applied in *Thunderer* and *Hurricane*. Details of how many were built plus dimensional and performance details have not survived but apparently at least one engine built under Harrison's patent was delivered to the Stanhope & Tyne Railway. Thus the design basics were rooted in practical experience as opposed to the group of 2-2-2s which were apparently contractor-derived attempts to meet a demanding parcel of specifications.

The novel approach adopted with the Harrison-patent pair may have been at Brunel's instigation in attempted circumvention of shortcomings in prevailing design practice. They embraced even more unusual features than the seventeen preceding locomotives. As an exercise in original thinking, they were the most extreme of the group which makes them the most interesting to describe and discuss. Certainly it seems likely that their inherent iconoclasm might have appealed to Brunel.

MacDermot was uncomplimentary, referring to the 2-2-2s as 'freaks'. He went on to label *Thunderer* and *Hurricane* as 'super-freaks' while *The Locomotive Magazine* described them as equally 'singular' and specifically referred to *Hurricane* as 'this monstrosity'. Notwithstanding their failure, the relationship continued as Brunel called in Harrison for consultative advice on matters pertaining to the Taff Vale Railway in 1841. The NER and GWR would maintain cordial relations over the years, the seeds of which might have been sewn through this collaboration.

WESTERN TIMES

The challenging weight constraint was nearly satisfied through separation of the boiler from the transmission assembly on different vehicles in an arrangement that might be considered an early form of articulation, notwithstanding the definition laid down by Professor Lionel Wiener in his standard work, *Articulated Locomotives* 1930 (re-published 1970 by Kalmbach Publishing Co, Milwaukee, WI). He stated that 'An articulated locomotive is a locomotive in which one or more of the driven axles are able to take up positions where they do not remain parallel to the others and may take angular positions in curves'. By adopting the less scientific definition that articulation is achieved where the mainframe structure that carries the complete power unit [excluding the tender] bends or pivots, then the pair were uniquely articulated among GWR locomotives. So far as can be determined, the leading dimensions were:

Locomotive	*Thunderer*	*Hurricane*
Wheel Arrangement	0-4-0	2-2-2
Driving Wheels		
- actual	6' 0"	10' 0"
- equivalent through gearing	16' 3"	n/a
Carrying Wheels		
- driving unit	n/a	4' 6"
- boiler unit	4' 0"/ 4' 6"	4' 0"/ 4' 6"
Boiler Pressure [lb/ sq in]	50	50
Cylinders	16" x 20"	16" x 20"
Boiler Barrel	8' 8½" x 3' 8"	8' 8½" x 3' 8"
Firebox		
- width	3' 8½"	3' 8½"
- length	5' 0"	5' 0"
- height	3' 11"	3' 11"
Heating Surfaces [sq ft]		
- tubes	517	517
- firebox	108	108
Grate [sq ft]	17	17
Weight (engine unit only)	11-12 tons	11-12 tons
Adhesive Weight	6 tons	6 tons

Of the nineteen contractor-built locomotives *Hurricane* most clearly demonstrated the conflicting conditions imposed upon the design process. The locomotive's overall weight [i.e. boiler plus engine units] is unknown but may have been incorrectly recorded anyway as an early example of the builder's practice to be less-than-honest concerning locomotive weight to evade constraints imposed by the civil engineer. The driving axle is estimated to have borne around 50% of the engine unit's all-up weight with the carrying axle below the cylinder block bearing the greater share of the remainder. This suggests inefficient distribution which would have eroded the weight on the driving axle. Visually, the whole ensemble implied limited haulage capacity.

There is little authoritative information about the performance characteristics of the earliest GWR engines. *The Locomotive Magazine* [April 1901] stated that *Hurricane*, driven by Richard Wilkinson of the Stockton & Darlington Railway and fired by John Thompson was reported to have covered the 22½ miles to Taplow in 16 minutes. This would have equated a start-to-stop average speed of 84 mph which as the publication pointed out was obviously inaccurate. However, this information provides a basis for belief that the machine was fast by the standards of the day although there was no record of a trailing load.

Another source was a report in *The Times* of a coroner's enquiry during the second half of October 1838 when Mr WD Field, aged about 19 years was grievously and fatally mutilated while working on the line about 1¼ miles from Shepherd's Bush station. Field, together with a Mr Reeves, was in charge of a team of about ten working to monitor track stability and measure rail deflection with the passage of the 5.00 pm Down service. At about 5.15 pm, a whistle was sounded and this was perceived to be an 'experimental train' approaching from the London direction, on the 'north line' i.e. wrong line working. This train comprised *Hurricane* (described as 'very powerful, the wheels being 10 feet in diameter') plus seven wagons each loaded with 2 tons of iron, and was due to go no further west than Wormwood Scrubs. Messrs Field and Reeves intended to travel back to London on the return working of this train. A witness stated 'A ball was hoisted on a high pole as a signal when the train was about half a mile distant'. There were several men aboard the experimental train which was approaching at an estimated 2 mph. On seeing men at work on the track, those aboard gestured to the track gang to stand clear which they did, to the north of the Up line.

What happened next is slightly unclear from the press report but with *Hurricane*'s train 'about three perches off', Field crossed the Up line to attend to measuring equipment set between that and the Down line. (A perch is an ancient unit of length, roughly equal to 16½ feet). Field, who had apparently failed to notice the approach of the 5.00pm regular service on the Down line, was called back by his colleagues. He attempted to return across the path of the experimental train but failed to make it and was run over by five of the seven trucks before he could be drawn clear, and he expired within two minutes. Although unstated, it is inferred that Hurricane was propelling rather that hauling its load.

The inquest excited considerable interest and many attended including Charles A Saunders, Company Secretary for the Great Western and Dr Dionysius Lardner, later of Box Tunnel fame. Messrs Field and Reeves, two of Dr Lardner's articled pupils, had been deputed by him to supervise the track testing. *The Times* reported that Saunders informed the inquest that '...the experiments were not being carried on from motives of curiosity, but for real business. Dr Lardner had for some time been employed in making the experiments as a matter of utility, for the purpose of ascertaining the deflection of the rails'.

In giving evidence, Lardner stated that '...the deceased had been an articled pupil to him for full two years, during the whole of which time he had been employed on railroads in making experiments. The deceased was therefore much used to engines and railways, and had taken observations on all the principal railways in the kingdom. He could not account for his negligence in placing himself in that dangerous situation, as he knew no one more acquainted with the necessity of caution.' The jury returned a unanimous verdict of accidental death.

There are several points to note:

1. Beyond *Hurricane* sounding its whistle on approach about half a mile distant, the reported warnings concerning the wrong line movement of the experimental train, and apparently also the service train were solely visual with no audible element.
2. Coincidence was a factor i.e. the concurrent arrival of both trains moving in the same direction on parallel tracks.
3. A recurrent feature of incidents where track personnel suffered death or injury was attention focussed on one moving train with failure to notice another in the near vicinity.
4. It seems that *Hurricane* and train were there for test purposes. This engine entered service in October 1838 and must have been brand new. It is unclear whether its presence was in connection with Lardner's programme or because it was undergoing preliminary assessment. There is no record of *Hurricane* hauling a revenue earning service, nor of *Thunderer* for that matter. Both engines ceased work in December 1839.
5. Although *Hurricane* was in no way the cause, its association with the fatality would have detracted from the locomotive's reputation.

Companies in their establishment/ fledgling phase may encounter many, sometimes completely unforeseen, risks that can damage their survival prospects, perhaps terminally. It can only be imagined what were Saunders's feelings at the inquest. The company's motive power fleet had thus far encountered numerous problems and significant modifications were in hand to attempt redress of shortcomings arising from the unworkable design specifications imposed upon the builders. Breakdowns were legion and the service timetable had all but collapsed. Whatever the causes and the individual circumstances, the company must by then have been in poor standing with the travelling public. Saunders might have feared that the tragedy could prove to be the last straw. However, there were grounds for hope in the company's twentieth locomotive, a 2-2-2 delivered by R. Stephenson & Co on 28 November 1837. This was a straightforward design, free of the conflicting elements that had hamstrung the preceding nineteen. Hope rested also in a 20-year old engineer who had been appointed Locomotive Superintendent on 18 August 1837. Despite prevailing difficulties, *North Star* and Daniel Gooch were about to become synonymous with the unfolding majesty of the Great Western locomotive epoch.

A hardback guide entitled *London* was published in June 1923 for attendees at the International Railway Congress. A one-page advertisement by R and W Hawthorn, Leslie & Co Ltd displaying three of the company's past products was included in this publication. Commercial manufacturers occasionally used advertising images that promoted the diversity of their productive capacity but it is unclear what benefit might have been expected to derive from depiction of an experimental, unsuccessful oddball locomotive built to an obsolete gauge.

Acknowledgements / Bibliography

Grateful thanks to Peter Rance of the Great Western Trust for the photographs of *Hurricane* in model form, and for the excerpt from *The Times*.

- ***History of the Great Western Railway Volume I 1833-63***. ET MacDermot & CR Clinker. Ian Allan 1964.
- ***The Locomotive Magazine***. April 1901.
- ***The Locomotives of the Great Western Railway Part 2***. Railway Correspondence & Travel Society 1952.
- ***The Oxford Companion to British Railway History***. 1997.

MODERN TRACTION:
DIESEL MULTIPLE UNITS IN COLOUR

Top: We begin our colour selection of Western Region diesel units in South Wales, with a scene still pervading the very essence of the 'old company' at Ebbw Vale on 2 April 1962.

Waiting patiently in the platform for the board to drop is a 3-car BR Derby Class 116 unit forming the 11.35am departure for Aberbeeg. This is one of the later 1958 built batch of these units, delivered with two-character headcode boxes and only two marker lights.

The first livery chosen for these high-density suburban units was an austere plain green, relieved only by the front end 'speed whiskers', red buffer beam and white cab roof (which soon became filthy in traffic). ***P Strong.***

Middle: An original 1957 built Class 116 unit, featuring the four marker lights, stands at Crumlin (Low Level). Opened on 23 December 1850, the station was to close on 30 April 1962, shortly after this photograph was taken.

The view is dominated by the iconic Crumlin Viaduct soaring above the valley floor. Opened in 1857 this huge wrought iron structure carried the Pontypool to Neath Line on a deck some 200 feet above the Western Valleys Line and River Ebbw below. It remained throughout its working life the tallest railway viaduct in the United Kingdom (and third in the world), until demolition in 1965. ***Gerald Daniels.***

Bottom: Another ubiquitous Class 116 unit stands at Aberbeeg with a train heading up the Western Valleys Line to Brynmawr. The station at Aberbeeg consisted of four platforms situated within the divergence vee with the Ebbw Vale branch, out of sight to the left.

Sixty-six of these units were allocated to the Cardiff area, for working services such as those depicted on this page. As delivered they featured cramped seating, no gangway connections between the vehicles, or toilet facilities; which made them unpopular in comparison with the loco hauled steam-era stock which they replaced. ***Gerald Daniels.***

Above: The GWR placed great stead in utilising single-car diesel units on branch lines, a tradition that continued post nationalisation. Here we see Associated Commercial (AC) Cars four-wheel railbus No. W79975 entering Yeovil Town in early 1964, one of four of this type of vehicle allocated to the BR(W) between 1958 and 1967 before transfer to Scotland. *Bernard Mills.*

Below: Pressed Steel built 'Bubble-Car' of Class 121 is seen at Yeoveney Halt on a West Drayton to Staines West branch service. Whilst the date or identity of the unit is not recorded, it is one of the nine allocated new to Southall from late 1960. *Gerald Daniels.*

Above: A veritable DMU convoy heads south from Dawlish Warren past Langstone Rock and out onto the sea wall on 16 July 1966. The formation is led by a Class 121 single-car unit, followed by a Class 116 three-car set. The three vehicles nearest the camera appear to be a hybrid of Class 119 Driving Motor Brake Composite and Trailer Buffet Second lavatory, with a Class 116 Driving Motor Brake Second bringing up the rear. *Bernard Mills.*

Below: Gloucester RC&W Cross-country Class 119 departs Dawlish station towards Kennaway Tunnel. The date is Saturday 30 July 1966 and the crowds of holidaymakers on the esplanade seem unconcerned with the historic event unfolding at Wembley Stadium in London, where England were to win the football World Cup. *Bernard Mills.*

Having just left Saltash station and curving sharply onto Brunel's famous Royal Albert Bridge, is Pressed Steel Class 121 single-car unit No. W55029. By the time of this April 1967 view, DMUs were well established on the local Plymouth area services, having replaced the much-loved 64xx hauled auto-trains. *Bernard*

Another view at Saltash, this time in the goods yard in August 1966, features two Plymouth Laira allocated DMUs. To the left is a later build 3-car BR Swindon Cross-Country Class 120, with four-character headcode box. The Class 116 on the right is preparing to pull into the station platform in the Up direction towards Plymouth. *Bernard Mills.*

It is high summer in the August of 1966, as a 2-car Birmingham RC&W Class 118 sits at Looe awaiting custom for a return trip up the branch to Liskeard. The Western Region ordered fifteen of these high-density units (of which three were delivered as 2-car sets) and all originally worked in the West Country. *Bernard Mills.*

Above: Probably the most eye-catching diesel multiple units to operate over Western Region metals were the Blue Pullman sets employed between 1960 and 1973. In summer 1965, an eight-coach South Wales working heads away from Paddington and past the locomotive servicing yard at Ranelagh Bridge, where 'Warship' No. D819 *Goliath* is seen stabled. *Bernard Mills.*

Below: It would be remiss to present any feature on BR(W) diesel units and not include an original GWR designed example. Withdrawn and awaiting their fate at Swindon in January 1963, twin-car W33 and W38 paint a sorry picture. *Roger Thornton.*

Above: The date is Friday 4 May 1973 and the final public Blue Pullman working on the Western Region prepares to depart from Platform 6 at Paddington. The set formed the 4.45pm service to Bristol Temple Meads, with power car W60092 leading the eight-car formation. No formal fanfare accompanied this historic event, although an enthusiast special did run the next day. ***Bernard Mills.***

Top: A three-car Class 116 draws to a stop at The Lakes Halt on the North Warwickshire line to embark passengers onto a Birmingham Moor Street to Henley-in-Arden service. The 1957 Derby built unit carries the short-lived corporate blue livery with small yellow warning panel, but still retains the attractive white cab roof, a legacy of the BR green days.

Delivered as high density suburban units, with no through gangways or toilet facilities, Tyseley housed twenty-nine sets for working in and around England's second city.
Gerald Daniels.

Middle: BR Derby built Class 115 four-car unit stands in the Up platform at High Wycombe on a stopping service to London Marylebone. Whilst these 1960 vintage DMUs were officially Midland Region allocated, they provided the backbone of the suburban fleet on the former Great Western/ Great Central Joint line.

In relative terms the station at High Wycombe has survived the passage of time, and whilst the through lines are now gone, many of the buildings and even some of the magnificent running in boards remain in situ. ***Gerald Daniels.***

Bottom: Pressed Steel Class 121 No. W55030 stands under the overall-roof of the branch terminus at Henley-on-Thames. Fifteen of these versatile single-car units were built for the Western Region, along with ten trailers designed to run in tandem on more heavily loaded turns. This example was delivered to Reading Diesel Depot in December 1960.

The station at Henley was opened by the Great Western Railway on 1 June 1857, originally with three sizable platforms to cater for the Royal Regatta traffic. This was reduced to the two platforms seen here in March 1969 and the original trainshed and main buildings were demolished in 1975.
Gerald Daniels.

RECOLLECTIONS OF A WESTERN APPRENTICE

BRIAN WHEELER

Brian Wheeler is a second generation railwayman, whose father Bertram Charles Wheeler was educated at Marlborough Grammar School and started work for the GWR as a junior clerk at Bedwyn in the early 1920s. He recalled seeing 'Star' class engines regularly working the Cornish Riviera Express, prior to this working being taken over by a 'Castle' class which therefore dates the period to pre 1923/ 4. Mr Wheeler (Snr) subsequently passed the GWR Accountancy and Signalling examinations, and likely also First Aid.

By WW2 he was working in the emergency control office that had been established at Didcot. (Was this an outstation in case the regular control office at Reading had been put out of action, or was it a satellite office to deal with the volume of traffic heading south via the DNS line to Southampton Docks?) He had volunteered for service in the Navy in WW2 but was refused due to being in a reserved occupation. It is known he also spent some time at the temporary railhead at Newbury racecourse.

Amongst his duties, representing the District Operating Superintendent at a local level, was liaison with groups and societies wanting to run special trains starting in the London Division. One of these society organisers was recalled as regularly pestering Mr Wheeler for footplate rides. He would regularly attend Reading cattle market to liaise with farmers and breeders on the running of specials trains/ vehicles for cattle, and similarly trainers with horse boxes/ horse box specials. His office was next to the Reading Control office behind the buildings on the up relief platform and so overlooking the signal works. In BR days he rose to the position of senior relief station master for the London Division where he covered for Paddington, Slough, Reading and Oxford, before retirement in the late 1960s.

Father and son did several rail-based tours including to Scotland, where they lodged with the local policeman at Mallaig who had a side-line in bed and breakfast. Also, to Ireland riding on the narrow gauge lines and nearer to home on the Southern.

The family home was at Newbury where as a child Brian would spend long periods standing by the railings at the western end of Station Road opposite Newbury West signal box. Thus, railways were clearly in Brian's make up in his earliest days, for as a four-year old he left home one day and made his way on and across public roads to the gates leading to Newbury goods yard to watch the trains go by. At age seven he recalled being taken on a footplate ride around the circuit of Newbury racecourse on a USA shunter. He also recalls a photograph shown to him by his father of a 38xx leaving the racecourse sidings with a train of 0-6-0 'USA Tank' engines.

In childhood he recalls an O gauge Hornby railway at home – later changed to 'Dublo' – and also the challenge faced in obtaining any new items and likewise Dinky cars in the immediate post-war years. At that time in Newbury there were a few franchised Hornby dealers, including W H Smith, although fortunately the family lived close to the home of the manager of another toy store who would 'tip the wink' when a delivery was expected hence Brian could go to the shop to pick out what item he might want.

Above: In June 1945 an unidentified 38xx prepares to leave the US Army Depot at Newbury Racecourse, with a train of five USA 0-6-0T engines. The individual locomotives cannot be identified from this image but another badly torn print has the numbers of the first two as 1390 and 1412. Neither of these ended up on the Southern Railway, so it is likely this particular consignment was destined for overseas shipment. ***Brian Wheeler collection.***

There was also the opportunity for footplate rides at Newbury, Didcot men being recalled as willing to give an enthusiastic youngster the opportunity to travel on the footplate from the down platform under Blackboy's Bridge and reverse back into the bay. These were Didcot – Southampton trains which had a layover at Newbury. Often a new 22xx might be in charge, perhaps No. 3210, 3211, or 3212, some of the drivers complaining that because they were so new the reversing gear was stiff to operate. The crews would also bring some cotton waste to allow for unofficial engine cleaning.

The locomotive exchanges of 1948 were also a good time for the young enthusiast. Brian and his friends often made their way to the 'new bridge' over the main line and up goods loop between Newbury station and Enborne Junction where they would stand to watch a Merchant Navy, Duchess or A4 working trains from the West Country.

As an older boy, he and a friend were invited to visit and later operate a 5 inch gauge ground level railway in the grounds of a large property at Penwood on the south side of Newbury. Appropriately this was known as the Penwood Model Railway and was owned by a well-to-do gentlemen by the name of Nigel Holder. The railway consisted of several well engineered models including a Furness 'Baltic' tank, an A3 and Princess Royal pacifics, plus a Southern L1. It was being extended at this time and one of the jobs enjoyably undertaken by the boys was laying the new track, payment for which came in the form of regular 'drinks and cake'. (Little information appears to have survived on this railway and any new detail would be welcomed).

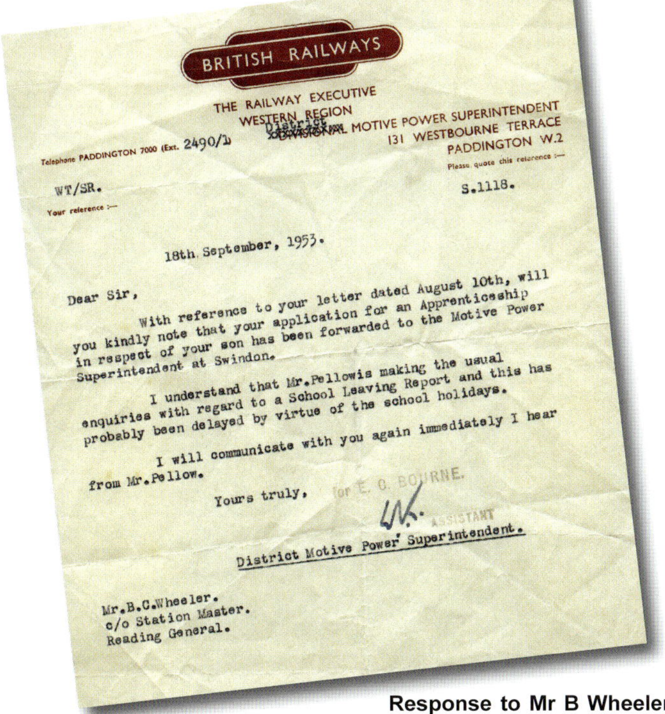

BRITISH RAILWAYS

THE RAILWAY EXECUTIVE
WESTERN REGION
District MOTIVE POWER SUPERINTENDENT
131 WESTBOURNE TERRACE
PADDINGTON W.2

Telephone PADDINGTON 7000 (Ext. 2490/1)

Please quote this reference :— S.1118.

WT/SR.
Your reference :—

18th September, 1953.

Dear Sir,

With reference to your letter dated August 10th, will you kindly note that your application for an Apprenticeship in respect of your son has been forwarded to the Motive Power Superintendent at Swindon.

I understand that Mr.Pellowis making the usual enquiries with regard to a School Leaving Report and this has probably been delayed by virtue of the school holidays.

I will communicate with you again immediately I hear from Mr.Pellow.

Yours truly,
for E. O. BOURNE.
ASSISTANT
District Motive Power Superintendent.

Mr.B.C.Wheeler.
c/o Station Master.
Reading General.

Response to Mr B Wheeler (Snr) letter requesting an apprenticeship place for Brian in the Motive Power Department.

Brian Wheeler is photographed driving the Furness 4-6-4T on the Penwood Miniature Railway. The railway featured at least once in the local *Newbury Weekly News* and was open to all on summer weekends. The track was to 5" gauge and ran for a quarter of mile. Its fate together that of the locomotives used is not known. ***Brian Wheeler collection.***

Brian had already made up his mind he wanted to work on the railway with the initial desire to be on the footplate. This was opposed by his father on the basis of unsocial working hours and limited promotion prospects. An alternative suggested was the Signal Works at Reading but this was not to Brian's liking and instead a compromise was reached with a fitter/ turner apprenticeship at Reading shed at the end of the 1953 school year. School had involved metalwork including turning, the teacher an ex Woolwich Arsenal foreman. The apprenticeship was for five years, four years at Reading, including college 'day-release' followed by a final year at Swindon. In his first year college time was during the evenings but following a family move to Reading, this became day release in the second year. College was not all theory either as it was here that Brian was taught further lathe and milling work.

The family were still living at Newbury at the time which meant Brian travelled to and from Reading by train. He recalls his first day as being on the 7.38am service from Newbury (originating from Hungerford) to Paddington. The engine turn for this was a Reading 'Hall' or 'Grange' (on his first day it was a 'Grange' – but the individual identity is not remembered). The engine would work light, chimney first to Newbury where it would turn at the racecourse; thence tender first to Hungerford to collect the passenger stock from the long siding behind the station on the up side. Once coupled up there was time to warm the carriages as necessary before pulling forward and then shunting back into the up platform in time for departure. It was a regular nine-coach rake of vehicles for this service; a single corridor coach followed by six suburban vehicles, sometimes articulated, and concluding with a further two corridor coaches. Using this service Brian's arrival time at Reading was such that that he started later than would normally be the case. Consequently, he was expected to work a Saturday morning to make up the hours.

The first day saw him walk out of the station turning right down Station Hill – passing the Eames model shop (which will feature later) – across the road, continuing through a park, under the railway and eventually up the slope to reach the GWR steam shed. Then it was over the nine shed roads at the end of which and facing the front of the shed was the shed master's office. On the first morning Brian knocked on the door of the office and was met by Percy Stroud the chief clerk. Bill Lloyd was the shed master whilst the mechanical foreman was Bill Miles, an ex MSWJ man. In conversation in 2022, Brian admits one of his regrets was not asking Bill Miles about his former life, even so the young apprentice must have made a good impression over time as Bill Miles was the instrument that led to several opportunities. Brian suspects this was because he was not just learning a trade but also showed interest in the railway. Other apprentices, of which there were five, were well capable but to them it was more of a job. (On his retirement, Bill Miles was replaced as Foreman by Vic Jarrett, who went by the nickname 'The Bishop' in consequence of his 'monk' style of hair. Vic had previously been the mechanical foreman at Oxford).

That first day saw Brian given his brass pay check and put with fitter Bob Gould and fitters mate Vic Harris, their task being to look around 'cold' engines on the washout road checking for any defects. If issues were found they were noted and the requisite work allocated accordingly. Defects identified and/or reported by a driver when an engine was in steam were similarly reported and dealt with. On that first day Brian also joined the Amalgamated Engineering Union and was issued with a blue card – this would be changed to a green card when he was time served. Bob Gould was also an official of the Reading branch of the AEU and was time served in an Aberdeen shipyard.

Staffing levels across all grades were good at Reading, there was even cleaning staff. Folklore has it that most steam sheds seemed to have an initiation ceremony for new cleaners but Brian recounts there was no such right of passage for the apprentices.

Above: Brian is captured giving some care and attention to the nameplate of Castle Class No. 5011 stabled on the long siding alongside No.9 road at Reading shed. He is wearing his trademark tie which he always wore to work, a discipline he maintains to this day!

Above: No. 6101 was a long time Reading engine. The image shows the Shedmaster's office door on extreme right. The two windows are the Running Foreman's office. The two steel pipes shown, one with a four-hole flange, supplied hot water to the Lifting Shop washing facilities and were fed by the stationary boiler, the latter situated alongside No. 9 road. The signals in the background were for the down main line. The Large Prairie is standing on No. 3 road. ***R C Riley (RCR 10564).***

For a portion of Brian's time Reading also had an allocation of a former GWR diesel railcar; either No. 18 or No. 19; this was used on the Lambourn branch. There was also one of the GWR 'twin' sets used on main line stopping services often to Newbury. These were serviced inside the steam shed on No. 1 road by a dedicated fitter who was a former AEC employee. The vehicles were recalled as generally reliable but due to the hard water at Reading the steam heat boilers were regularly flushed through with an acid solution and hand pump to remove scale. The diesel fitter had as his office a building erected post war and originally part of the abortive oil-burning scheme; Reading being one of the depots that had been provided with oil refuelling facilities. Several of the skilled fitting/ maintenance staff were time served in industries other than railway and included a former ship yard worker. There was also a specialist fitter, Dave Russell, who would deal with injectors, clacks and the like whilst two other fitters were specially trained to service the railcars. So far as boiler/ firebox condition was concerned, these were examined locally whilst in addition they were further inspected by a travelling boiler inspector.

When asked about his views on engines originally designed for other railways, he naturally admitted each railway's staff would no doubt prefer their own machines although one class was definitely disliked and that was the WD class of 2-8-0.

Brian does not recall any exchange of maintenance staff between the WR and SR depots at Reading. WR engines diagrammed to work trains from Reading (Southern) to Redhill were maintained and prepared at the WR shed and then made their way light to Reading Southern where they would back on to their coaches. On one occasion, the circumstances not recalled, a Southern Mogul arrived on the WR shed requiring attention, which unsurprisingly involved a part not held in stock. The easiest option was to walk over to the Southern shed which Brian did with one of the fitters. The part was made available to them but Brian's particular memory was the smell inside the shed – both steam depots of course but the smell within the Southern premises was so different to what he was used to. As he put it, '...perhaps they used a different type of oil...'.

Part of his learning also involved going out on derailments with the gang and Bill Miles, whom he described as excellent on breakdown work. He recalls two incidents both very different; the first in Pangbourne goods yard where a coal wagon had come of the rails, as he put it, '...nothing particularly difficult...it probably went back on with jacks'. The second one though was different and involved a Hall class engine off 'all-wheels (engine and tender) at Henley. Bill Miles decided the best way was simply to drive it out and so ramps were placed and the driver instructed 'to give it the works' – which he did – and it worked. As Bill put it there would be a few (more) broken chairs and fishplates but the engine was back on in double quick time. Immediately after the first task was to gauge the wheels to make sure the gauging had not been the cause of the derailment. All was in order and although the engine had not been able to take up its next booked working it was back on the rails and available to traffic quickly. The original cause of the derailment was not mentioned.

It was through Bill Miles that Brian was first informed of and later given the opportunity to participate in some special workings. One of these was the 1954 special working of No. 6003 *King George IV* on the prelude to what would be the down 'Bristolian' due to be restored to its 105 minute schedule. To test the timings the speed limit through what was then Platform 4 at Reading (the down main) had been especially raised from 60 to 75mph and Brian was there to witness No. 6003 come roaring through complete with the WR General Manager Mr K W C Grand on the footplate. As is well known the trial was a success and the revised schedule was implemented. Later it became almost a ritual for Brian and some of his colleagues to witness the passing of the down train at 9.18am each morning so much so that if anyone enquired where he was around that time the answer would be '....watching the train'. One incident involving a renowned service was when the up 'Cornish Riviera Express' which was brought to a dead stop at the bracket signal on the curve from Reading West as the signalman had allowed a Southern T9 to depart towards the station to work a Didcot service, but the engine could not find its feet and kept slipping so preventing it getting away.

Mogul No. 6324 was a Reading based engine during Brian's time and he remembers that together with apprentice Alan Brown they noticed that on one side of the tender in particular the words 'GWR' could still be seen although by this time starting to fade. Old standards clearly to be maintained and consequently the two of them visited Eames to buy paint and brushes after which they proceeded to line out and restore the wording on the fireman's side of the engine.

Other incidents that were what we would now regard as good for morale included when No. 5018 *St Mawes Castle* was specially prepared as the stand-by engine for a Royal working due to pass the station, which included painting the guard irons and other items that were more usually left black with white paint.

Another Mogul, No. 6385 also received something that may have been unique for a WR engine, this was a blue background to the smokebox numberplate. At the time the engine was in for attention in the fitting shop and Brian thought it would be a good idea if the background appeared as per some of the Scottish Region engines which were being similarly painted around this time – Eames supplies again. How long it lasted as such is not reported. Some of the Moguls and also the Manor class engines working between Reading and Redhill also had the cab backheads scoured to perfection by their crews such was the pride in the job that existed at that time.

A once in a lifetime experience came to Brian in November 1954 when again through the kindness of Bill Miles he was invited to travel in the dynamometer car of the 20-coach controlled road test of County No. 1009 *County of Carmarthen* which had recently been fitted with revised draughting arrangements. The engine and car worked light to Reading from Swindon where it was turned on the triangle and attached to a rake of vehicles assigned by the Carriage & Wagon foreman at Reading. As Brian puts it, "I rode in the dynamometer car. The chiefs from Swindon were all there including Mr Ell and Mr Dymond. I was introduced and perhaps because I was interested, was shown how recordings were made and the results being achieved. At Stoke Gifford we turned on the triangle and came back to Reading. It was fascinating. The 'chiefs' all seemed pleased with the results which was as a result of improved draughting being fitted. I was even getting paid for the day as well".

An engine regularly seen on the 'Bristolian' was No. 4090 *Dorchester Castle*. Along with No. 4093 *Dunster Castle,* both had been fitted with a 4-row superheater in 1957 and an associated slightly longer smokebox. They were among Old Oak Common's star performers and consequently were frequently rostered to haul this working.

Above: When Brian started at Reading in October 1953, the MSWJR 2-4-0s previously allocated to the depot had already departed for scrap. However this former MSWJR tender remained into the 1960s, with identification of two of its previous owners still showing. The tender plate was subsequently liberated by Brian but is no longer in his possession. ***Roger Holmes.***

Above: County Class 4-6-0 No. 1009 *County of Carmarthen* at Swindon on the trials referred to in the text. As a young apprentice Brian travelled in the dynamometer car on these runs and recalls speaking to Mr Ell and other senior officials of the CME's department. He recalls it as a memorable experience. The test train is seen here on its outward run at Swindon in November 1954.

Below: A view of the test chimney fitted to No. 1009 recorded on another occasion at Swindon Works. The chalk writing 'Brakes wanted off 5081 (*Lockheed Hudson*) please' is not explained. *R H G Simpson / Transport Treasury.*

Another incident I recall was a few years earlier when our Dean Goods No. 2474 was rostered to work an enthusiast's special to Shipstone-on-Stour, the tour of 24 April 1955. The engine was checked over but was found to be blowing steam from where the dome sat on to the boiler. This was a job for Dave Russell the man who looked after injectors etc and accordingly the asbestos ring that sat between bottom of the dome and the boiler was replaced. For whatever reason steam continued to blow and this may be seen in several photographs of the engine at work on that day. Dave was somewhat embarrassed about the affair. The spare had been sourced from the stores at Reading although Dave admitted it was an unusual repair as it was normally the seating ring between the safety valves and the boiler / firebox that would need replacing. Brian also replaced the round spectacle glass on the now preserved No. 2516 but at the time it was still in everyday service. The fireman's side glass having a crack the next question was would the stores have a glass in stock? They did and it was but a quick task to take apart the brass surround and replace the glass. It is perfectly possible Brian's handiwork still remains on the engine in its preserved condition.

Above: No. 2474 on special duty at Shipstone-on-Stour on 24 April 1955. The image certainly records the dome joint blowing well - perhaps not Reading fitter Dave Russell's finest hour! ***R C Riley (RCR 6078).***

After four years at Reading, Brian moved to Swindon for his final year as an apprentice where he lodged with a Mrs Smythe in Shrivenham Road close to the County Ground. In the same house was a young man from Wellington (Somerset), who worked as a clerk in one of the stores offices and who was also an enthusiast. The work at Swindon was more specialised than at shed level for obvious reasons. The first period of time was spent jig-drilling boilers that had come from the boiler shop with new smokeboxes. This involved drilling holes for the saddle holding down bolts and likewise for the steampipe apertures and lance cock fitting etc. Another period of time was spent on fitting various items to boilers including safety valves and clack boxes plus assorted backhead fittings. These were water pressure tested. Being with a fitter assigned to the diesel section who also covered the gas turbine No. 18000, Brian was given opportunities others may not have had. This included two runs on No. 18000 after repair. Both were light engine, the first the short distance to Kemble and the second to Hullavington. Unfortunately, at the latter location the engine broke down and consequently they were towed back. He recalls it was not the most popular machine but partly no doubt due to being a one-off. There was also time spent in the diesel test house 'The Barn' on 350hp shunters. He recalls having a mini tour of the carriage works located on the down side opposite the loco works plus a trip to what was referred to as the 'Crystal Palace' where name and number plates from scrapped engines were stored. He recalls also being told by the Chief Clerk at Swindon that he would also now have to attend Swindon technical college instead of Reading; fortunately, common sense prevailed and for his final year day release continued at his home town. (To prove attendance the apprentice took with him a card which would require to be signed by the college. He would then bring this back in to work to confirm he had been to the college.)

Completing his apprenticeship, the next move was away from the railway and into the army for National Service. Railway staff so conscripted would have their job held open provided they only completed the requisite minimum of two years, any more than that and it was considered you had resigned. Accordingly, Brian's first choice to go to the Longmoor military line had to be rejected as this was only available on a three year posting. Instead he chose an admin role, something he says he did not regret for what occurred in the future – including meeting a lady called Barbara, a nurse in the hospital at Osnabruck whilst posted to Germany who later became his wife and mother of his two sons. She was destined to be very supportive of his later work in preservation.

Brian was demobbed and returned to the Western Region at Reading in 1960. A lot had happened in the ensuing two years since he had last been at the depot, notably the widespread use of diesel multiple units. He was allocated a position within the diesel depot but admits he disliked it immensely and after just two weeks requested a return to work on steam. As a time served fitter, now it was his turn to have a mate – and sometimes an apprentice. Brian kept copious notes on the engines he worked on during that time which included examples from most of the steam classes still operating including No. 5068 *Beverston Castle* which was towed to Reading as the inside motion had failed at Twyford, the engine working a Fishguard Harbour to Paddington service on 2 August 1961. At the shed the engine was examined and placed on No. 8 road within the shed. Here the inside motion was dismantled, repairs effected and it was returned to service after 10 days. Brian's records include work on No. 4700 plus other Castles; details of piston and valve sizes recorded by using a micrometer and written in a neat hand in a small hardback book, the cover of which was originally blue but which remains heavily stained with oil decades later.

Knowing of Brian's enthusiasm, Bill Miles arranged for a visit to Didcot to see No. 3440 *City of Truro*. After its Didcot allocation, the celebrity locomotive was moved to Reading where it continued to be called upon for enthusiast specials. Brian admits it was perhaps not quite as cherished as when it had been at Didcot, a point that was proven when it was alongside the Reading coaling stage and a pile of ashes on the ground alongside the tender burst into flames scorching the paintwork. The engine and tender had to be returned to Swindon for repainting. This scene at Didcot with a host of shedmates. *Brian Wheeler collection.*

Above: No. 5068 *Beverston Castle* in for repair. Left to right: Brian Wheeler, fitter, (always wore a tie both in the shed and works), Alan Brown, apprentice fitter (came from Newbury) and John McNamara, apprentice fitter (came from Didcot where his father was a Driver). Photo by the late Bob Judge a fitter at the time and who originated from Gateshead Depot. The Castle is standing on the blocks of No. 8 road, whilst the engine to the right No. 4957 *Postlip Hall* was on No. 7 road for a boiler washout. *Brian Wheeler collection.*

Conscious however of the limited time left for steam, he applied for and was successfully appointed on to a supervisory course starting on 19 March 1962. This was for 63-weeks duration and was notionally based at Paddington although on-site experience was a considerable aspect of the course. In consequence he was 'allocated' to Didcot where the shedmaster was George East and the mechanical foreman was Arthur Brinkly. Time was spent with various members of staff including the shedmaster, chief clerk, general clerical store keepers, running foreman, mechanical foreman and lastly boilersmiths. He was also issued with a footplate pass and recalls several trips from that time, many with Didcot crews. One example involved running light from Didcot to Oxford with No. 6350 and Didcot driver Dodds to pick up a freight for Eastleigh via Newbury; '39 with 15 as a fitted head'. This train was taken as far as Eastleigh where they relieved Southern men on a 9F in charge of a Fawley – Bromford Bridge oil train again over the same route and were in turn relieved at Didcot North junction. Brian recalls firing the 9F between Winchester and Enborne Junction.

Other examples included on 19 April 1962 with No. 1002 *County of Berks*, on an ECS trip from Didcot to Reading where the engine was then taken to shed and turned. It then worked the 2.30pm from Reading to Swansea '11 equal to 354 tons'. He notes they were three minutes late departing, the engine was rough to ride on and that the exhaust injector was not working when the regulator was fully open. Despite a permanent way check at Shrivenham some time was regained and the arrival at Swindon was just two minutes down.

Left: At Reading shed, No. 1000 *County of Middlesex* on the line which ran the full length of the shed building and could be reached either from the coal stage inlet or from the points which led into the lifting shop road. As can be seen, both sets of valves have been removed. Adjacent is the two wheeled 'Iron Horse' used for moving connecting and coupling rods into the lift shop when required for attention such as re-metalling bushes. *Eric Sawford.*

Below: No. 1442 standing outside the Reading Lifting Shop which housed the forge, where on a cold day Brian might go to warm by the fire; smoke from his oily overalls being not uncommon. The brick building just visible between the smokeboxes of the two locos is the Mechanical and Boilersmith Foreman's Office. It was close to here that every morning Brian would stand to watch the down 'Bristolian'. *Eric Sawford.*

Above: Reading pannier tank stalwarts, Nos. 9403 and 7777 stand in front of the main running shed, whilst an unidentified AEC Railcar in carmine and cream livery lurks under cover. *Eric Sawford.*

He admits the highlight of his many footplate trips was probably on 12 May 1962 when he travelled on the footplate of No. 6021 *King Richard II* working the 9.10am down from Paddington to Wolverhampton '12 equal to 420 tons'. On the return run the engine achieved 92mph at Denham. This particular trip was outside his official training course but Brian had managed to convince the Old Oak Common shedmaster, Mr Simms, of his bona-fides and the trip was thus easily approved. At the opposite end of the scale Brian was on the pannier tank used to shunt the goods yard at Wallingford. Upon arrival the driver climbed down and walked off – where to was never established – but he went off simply stating the two of them (fireman and Brian) could cope perfectly well with the shunting without him.

Another trip involved travelling on the S&D south on the Pines Express with Branksome driver Bert Brewer who had the additional role of being the ASLEF rep at Branksome shed. When Bert learned of Brian's position he quickly turned, as Brian puts it, '…like the third world war had broken out…' adding to the footplate temperature on the Standard 5 which was being piloted as far as Evercreech by another Standard 5. It transpired that Brian had been seen as an emissary of the 'dreaded' Western Region with the latter's avowed intend on closure of the S&D – which of course they later did. Even so it was not long before the mutual love of the steam engine came to the surface and Brian recounts it was a memorable trip, latterly for all the right reasons.

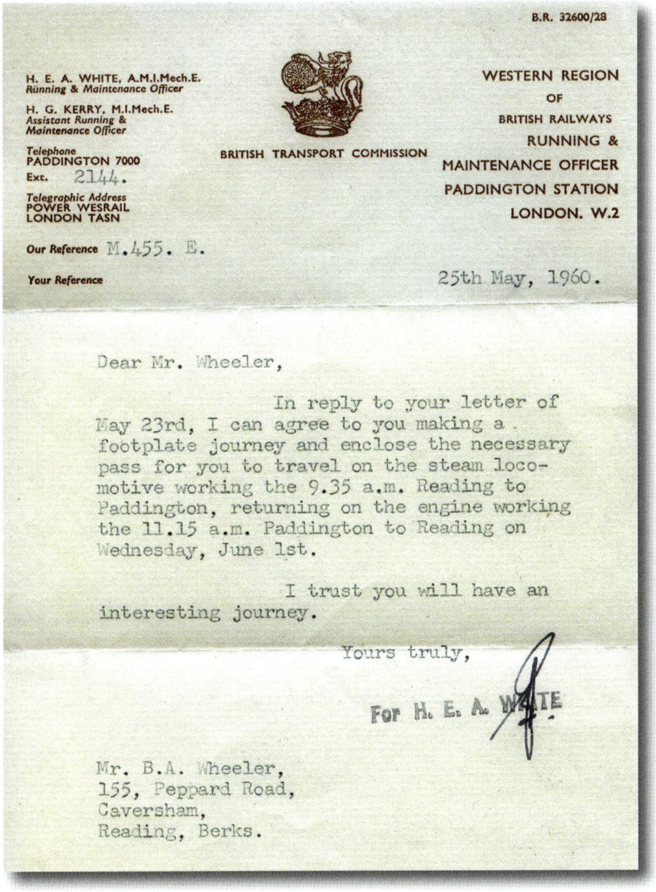

B.R. 32600/28

H. E. A. WHITE, A.M.I.Mech.E.
Running & Maintenance Officer

H. G. KERRY, M.I.Mech.E.
Assistant Running & Maintenance Officer

Telephone
PADDINGTON 7000
Ext. 2144.

Telegraphic Address
POWER WESRAIL
LONDON TASN

BRITISH TRANSPORT COMMISSION

WESTERN REGION
OF
BRITISH RAILWAYS
RUNNING &
MAINTENANCE OFFICER
PADDINGTON STATION
LONDON. W.2

Our Reference M.455. E.

Your Reference

25th May, 1960.

Dear Mr. Wheeler,

In reply to your letter of May 23rd, I can agree to you making a footplate journey and enclose the necessary pass for you to travel on the steam locomotive working the 9.35 a.m. Reading to Paddington, returning on the engine working the 11.15 a.m. Paddington to Reading on Wednesday, June 1st.

I trust you will have an interesting journey.

Yours truly,

For H. E. A. WHITE

Mr. B.A. Wheeler,
155, Peppard Road,
Caversham,
Reading, Berks.

Other aspects of the course included observing the work in Maidenhead East signal box, shadowing the Stationmaster for three weeks again at Maidenhead commencing 9 September 1963, and later a similar role seconded to the Yardmaster at Reading. In addition, he was assigned to Laira for training on main line diesel locomotives covering both the mechanical and electrical trades. Mechanical training was for four weeks and electrical a further two weeks. During this time, he lodged in the Laira hostel for the princely sum of 2/- per night, He would travel home at weekends. The time spend in Devon was from 23 July though to 1 September 1962 and when steam was still in evidence. Several engines were recalled from this time including Nos. 6000, 4037 and 4087 working an up seven coach ocean liner special.

The supervisory course eventually concluded and Brian was posted to the role of temporary mechanical foreman at Old Oak Common. This was at the shed, the repair shop at Old Oak was regarded as a separate entity. There was still some steam but main line trains were mainly in the hands of the diesels. Asked about his opinion of the 'new' motive power Brian was forthright, '…it was all done in hurry and what a waste of money. There were terrible reliability problems with some.' He also agreed some of this was simply down to lack of familiarity of both fitters and footplate crew. Completion of the course meant the individual was also expected to advance himself by perusing the weekly vacancy sheets for anything that might be appropriate. The vacancy sheets covering the whole of British Railways and consequently all regions. One job that did appeal was that of a shift maintenance foreman depot at Laira, he applied but was unsuccessful.

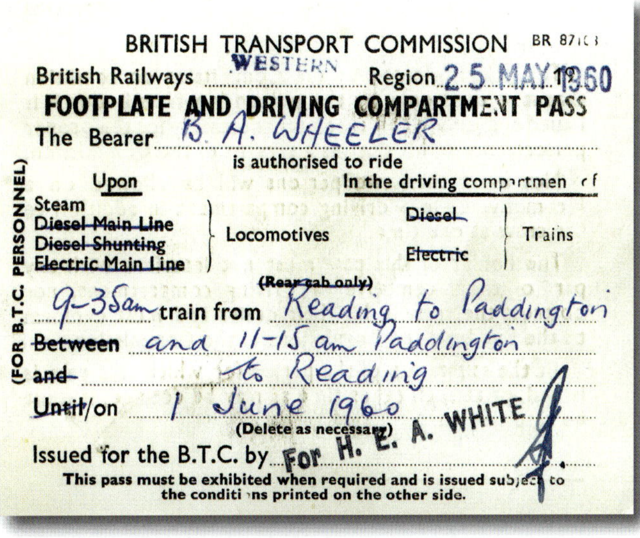

BRITISH TRANSPORT COMMISSION BR 87103

British Railways WESTERN Region 25 MAY 1960

FOOTPLATE AND DRIVING COMPARTMENT PASS

The Bearer B. A. WHEELER

is authorised to ride

Upon: Steam ~~Diesel Main Line~~ ~~Diesel Shunting~~ ~~Electric Main Line~~ Locomotives

In the driving compartment of: ~~Diesel~~ ~~Electric~~ Trains

~~(Rear cab only)~~

9-35am train from Reading to Paddington

~~Between~~ and 11-15 am Paddington

~~and~~ to Reading

~~Until~~/on 1 June 1960

(Delete as necessary)

Issued for the B.T.C. by For H. E. A. WHITE

This pass must be exhibited when required and is issued subject to the conditions printed on the other side.

(FOR B.T.C. PERSONNEL)

Whilst on leave from military service in Germany, Brian was very keen to further his railway experiences and consequently secured a number of footplate trips as described. The letter of authorisation from the Running and Maintenance Officer and the accompanying 'Footplate Pass', were prerequisites of these journeys. He recalls being accompanied onto the station platform by the Reading shed foreman Vic Smith who then left him to it. There was no inspector on the footplate.

He was however successful in his next attempt at the new Cardiff Canton diesel depot and as he recalls, was interviewed at Marland House Cardiff and was duly appointed. His spell at Cardiff would last for six and half years during which time the depot would, if necessary, even undertake the change of a diesel engine. In this aspect Canton might appear not to have been the fastest taking 48hrs for the task whereas Laira would manage the same job in half the time. A little investigation soon found the reason, Laira would physically cut the various hoses that formed connections to the engine whereas at Canton these were carefully disconnected and so could be re-used.

At that time the Kings had gone and the new diesel depot was in the course of construction. Main line and diesel shunters were serviced at the Swansea end of the DMU shed until facilities became available. To operate the passenger services Hymek and English Electric type 3s were allocated to the depot although individually neither could pull the same loads as the Kings they had replaced. Consequently, recourse was made to double heading with the diesels – not mixed types of course. (The D68xx type were also used in multiple for a time on the 'Bristolian' - see note).

Brian's spell at Cardiff was followed by a move off region to Liverpool Street to work in maintenance control, but he subsequently returned to the WR at Paddington. When interviewed he commented one of his regrets is that his father did not survive to see him return to the WR. His final move coincided with relocation of the Paddington office 77 miles down the line to Swindon, where he continued in maintenance control at 125 House. Brian retired in March 1993 and now spends his time perfecting his superb O gauge model railway based on Barnstaple Victoria Road – in GWR days of course!

Editor's Note: *Through the kind offices of Graham Howell using information from Mike Woodhouse and Alan Whitfield, we are able to give some more detail on the Bristolian run with Nos D6881 and D6882 which took place on 22 May 1965. These two machines were allocated to Canton but looked after on a day to day basis at Old Oak Common. The pair were equipped with higher ratio traction motor gearing. As in all English Electric type 3s, they only had two stages of field diverts and so suffered with poor high speed running.*

The pair worked the 8.45am Paddington – Bristol returning with the 11.35am Weston – Paddington with the 'XP64' stock. On 3 June 1965 they worked a special high speed Paddington – Plymouth trial but this proved to be unsuccessful as they performed below par on the Devon banks. A further high speed run took place in 1966 from Paddington to Cardiff and return which was better as most of the running was relatively flat.

In mid 1966 they were replaced on the Bristolian by No. D0280 Falcon *(later 1200) which was a success as this loco did not off-load at 95mph unlike the Brush type 4s.* Falcon *would achieve 100mph with no problem.*

Above: Modern times – well post steam anyway. The pairing of Type 3s Nos. D6881 and D6882 on 'The Bristolian' service. Due to its speed and available braking distances at some locations, operationally this train required signalmen to have two block sections clear ahead before 'Line Clear' could be given.

No. 4086 *Builth Castle* at Reading on 7 August 1955. Brian's notebook refers to an entry relating to identical sister engine No. 4085 *Berkeley Castle*, which arrived on shed on Christmas Eve 1961 with the left hand inside piston valve blowing. As the engine had been built with 'joggle' frames, it allowed sacking to be placed on the bogie frame and for Brian to crawl in. Tools and materials were then handed in through the frame cut-out by apprentice Alan Brown. The engine was still in steam and the repair took all of 30 minutes. If just a fitter and mate were on such a job, a regulator stop would be fitted which prevented the regulator being opened. *Eric Sawford.*

Souvenir from Swindon. Brian recalls his father arranged to collect the plate at the pedestrian tunnel entrance to the Works, and admitted being exhausted carrying it back to the station!

In preparation – The footplate riding experiences of Brian Wheeler in 1962 as an official observer. Part of Brian's formal training was to gain experience on a variety of services and motive power; including, King, Castle, County, Hall, Western, Warship, and Hymek classes. As he describes it, "There is nothing like being on the footplate of a King when the regulator is opened and the engine digs in to lift over 400 tons away smoothly….or the noise at 90 mph".

Above: Taken at Old Oak Common, No. 6007 *King William III* is in the 'factory' being prepared for a lift. John Wood (left) is supervising. *Brian Wheeler collection.*

READING SAND VAN

The pre-rationalisation railway provided an endless source of interest in the variety of outdoor furniture that helped meet operational needs including vehicles whose function could be obscure. Sand Vans were an example which spent most of their careers well away from the public eye. Photographs which have recently come the way of the editorial team depict a sand van standing in Reading West yard, and the background and purpose of this vehicle requires some explanation.

The largest single source of revenue for the Big Four was the transportation of freight and minerals, and well into BR days, the bulk of this traffic was moved in primitive four-wheeled wagons. Many trains were loose-coupled without continuous brakes, a feature that allowed small locomotives to haul prodigious loads. On starting a train, wagons were picked up one at a time and the gathering momentum made assumption of the increasing load progressively easier. Once on the move, that same momentum was no longer an aid but a major challenge when bringing the consist to a safe halt. Freight operations under these conditions were a subject too complex to explore in depth here but suffice to say that safe working called for extraordinary levels of skill and careful co-operation between locomotive crew and guard.

GWR locomotives had good brakes – as proven by the competence of the Buffalo 0-6-0ST as one of the company's principal freight types during the late 19th Century – but often insufficient to keep full control at the rear end of the train. Thus the guard in his Toad had to work his brake to judicious and timely degree, and to be watchful that insufficient friction between wheel and rail did not allow the former to lock with resultant loss of rear end brake power. Thus while sand might be needed to help a locomotive start, sand discharged from boxes in the Toad could be essential for the process of stopping.

It is believed that sand used at Reading was supplied from a site in South Wales. The normal process was to deliver it to the shed where it was dried in the sand furnace sited close to where locomotives were routinely serviced and replenishment of their sand boxes formed part of that process. Toads in between duties were parked at locations remote from sheds so there was a need to move sand to, and store it at, a location nearby. Enter the Sand Van.

Images courtesy *Ted West/ Mike King collection*.
Thanks also to Brian Wheeler, ex-steam fitter at Reading.

The van depicted is a Diagram V6 Iron Mink. The term 'Iron' was unofficial as the company built covered wagons with bodies in this material and timber-bodied vehicles concurrently (both code-named MINK or MINK A). Despite the unofficial title, components (door frames and floors) were actually timber. It is surmised that the door frame might have been the weak point in what was otherwise a tough construction leading to timber replacement doors as in this case. Five vertical external planks formed the standard design, reinforced by ten horizontal planks internally. These doors were slightly wider than the iron originals which necessitated relocation of the door retainers further apart on the bodies rather than on the vertical stanchions.

Externally, the appearance is entirely conventional except for the labelling and for the sand discharge pipe detailed in the second photograph. Presumably buckets were filled by opening a valve inside the van (details of the internal arrangements would be welcome) and then carried to the Toad needing a sand refill. The 'restricted use' nature of the van's duties are evident in the white diagonal cross on the right-hand door, the legend in traditional GWR script 'Not to run more than 3 Miles on the Main Lines', and the clear delineation on the left 'SAND VAN READING WEST JUNC'.

Diagram V6 was under construction from 1886 until 1901 and ultimately totalled 4901 wagons. Of this total, eight were built new for the Locomotive Department, ten for use in passenger trains, fifteen as stores vans. Iron Minks were later favoured for a variety of internal uses defined as: Compressor, Loco Stores, Loco Tool, Mobile Workshop, Sand, Sawdust, Signal Oil, Tool, Tool & Packing, Tunnel Centre, Tunnel Whitewash.

Iron Mink No. 59532 was built as part of (GWR) Lot No. 155, was transferred to service use as No. 35374 in February 1946 and allocated to Reading West Junction. It is recorded as having replaced another sand van which had carried the same number. This practice was similar to that with Shunters' Trucks which were allocated to specific depots and when a replacement was provided, it assumed the number carried by its predecessor.

Unfitted freight trains outlived steam on BR(W) with the services of Sand Van No. 35374 possibly remaining on this humble work. It may be speculated that it might have been one of the last surviving vehicles with genuine GWR lettering, and thus makes an interesting prototype for historians and modellers alike.

Above and Below: Neither of the archives' two photographs of Barry Railway Tri-composites record the complete coach side so both are included here for reference purposes. Multi class passenger accommodation (Tri-composite/ First & Second Composite/ Second & Third Composite) had appeared in four- and six-wheeled form previously but these were the only bogie examples, and at 52' 6", the company's longest coaches. The images show (upper) No. 6130 [156] at Cardiff Queen Street on 6 August 1951 and (lower) No. 6126 [151] at Caerphilly on 11 May 1952. Reference to the lower image indicates that as built the accommodation was: x1 third class compartment; x3 first class; x4 second class. Details of the seating provided by this arrangement is unconfirmed but would appear therefore to have been First Class (24), Second Class (40), Third Class (10). The GWR appeared uncertain how to categorise these coaches. Nos. 151 & 156 were down-rated the All Third in June 1923 and renumbered 4366 & 4364 respectively. Then in 1924 they were restored to composite status. In July, old No. 151 was re-designated First/ Third and assumed its originally allocated (but not carried) GWR number 6126. Old No. 156 followed the same course in September becoming No. 6130. They were withdrawn No. 6126 [151] – April 1959 and No. 6130 [156] – November 1957. *R C Riley (RCR 3394/ 3479).*

ABSORBED WELSH COMPANY COACHES

Those whose trainspotting memories go back to the 1950s may nostalgically recall the Isle of Wight as a custodian of rolling stock that dated from pre-Grouping years. Travel in an ordinary BR service train on the island was a delightful time warp that naturally suited the sylvan countryside. Other corners of the country could offer the experience of the nationalised network still operating a working museum and while not to the same exclusive degree, it was possible to sample passenger travel in the mode of earlier days. Wales was such a locality in retaining both locomotives and coaches that dated back to pre-Great Western ownership and which had remained on their native turf.

The greater proportion of the Transport Treasury archive is devoted to individual locomotives and to views of approaching trains while studies of individual coaches are few. However, during visits to Wales in 1951/2 Dick Riley fortunately recognised that the principality was the last home for a number of ancient coaches of Great Western origin, and also from the Welsh companies which had been absorbed in 1923. This pictorial selection profiles some of the surviving vehicles he encountered during those trips.

Coaches in this article are identified by their GWR running number, excluding the BR(W) applied 'W' prefixes and suffixes. The absorbed company running number appears after the GWR number in brackets.

Above: Apart from the 8-wheeled 47' Director's Saloon built by Cravens Ltd in 1899 (appropriately numbered 1), the entire coach fleet of the Barry Railway was four or six wheeled until a major advance in 1920. That year saw introduction of fourteen bogie coaches, built by Birmingham Carriage & Wagon Co. They were supplied in three types: All Third [six examples]; Tri-composite [four]; Brake Third [four]. No. 274 [199] was pictured on 5 May 1951 at (appropriately) Barry; it was withdrawn 31 January 1959. This type accommodated 50 passengers in five Third Class compartments with about 40% of its 51' length given over to space for the guard and luggage.

This was probably the best known Barry coach type as Nos. 263 & 268 [197 &198] were transferred to Devon in April/ May 1949 to provide passenger and guard accommodation for mixed trains over the Culm Valley branch from Tiverton Junction to Hemyock. They had been built with electric lighting but were retro-fitted with gas for service in Devon as speeds over the branch were judged too slow effectively to charge batteries. They were withdrawn in December 1962. ***R C Riley (RCR 3104).***

Above: Ex-Rhymney Railway 47' 10" All Third No. 1065 [40] is seen at Rhymney on 2 May 1951. At least 14 of this type were built, mainly by Gloucester Carriage & Wagon Co. However, when spare capacity at Caerphilly permitted at least five were built in house. The first appeared as early as 1900 but the example depicted dated from 1920. Internal layout comprised two open saloons, divided by a full partition. The saloon lengths differed with one containing four transverse seating bays and the other five, reportedly accommodating 76 passengers although this figure may have varied. Access to each saloon was by means of two doors on either side. This example remained in service until 26 January 1957. Some (but not this coach) were converted by the Rhymney Railway to work as intermediate non-driving trailers in auto trains, paired with the modified Brake Thirds seen below. *R C Riley (RCR 3072).*

Below: Another product of Gloucester Carriage & Wagon Co for the Rhymney Railway was this 47' 10" All Third No. 1174 [98], seen at Ystrad Mynach on 3 August 1951. This had the conventional eight compartment layout which provided seating for 80. Introduced in 1900, this coach was in service until 8 August 1953. *R C Riley (RCR 3354).*

Above: Brake Third No. 1176 [25] built by the Rhymney Railway at Caerphilly. While apparently open saloon type layouts were favoured, this coach appeared in 1915 with seating for 50 in five compartments. It was equipped in 1919 to work in an auto train whereby a driving cab was installed in the van end (All Thirds of that vintage were fitted to work as intermediate auto trailers). In August 1926, the GWR system of auto control equipment was substituted for operational compatibility in push-pull services but this equipment was removed in November 1934. Thereafter it worked as a conventional coach as seen here at Rhymney on 10 May 1952. It was withdrawn in December 1958. *R C Riley (RCR 3470).*

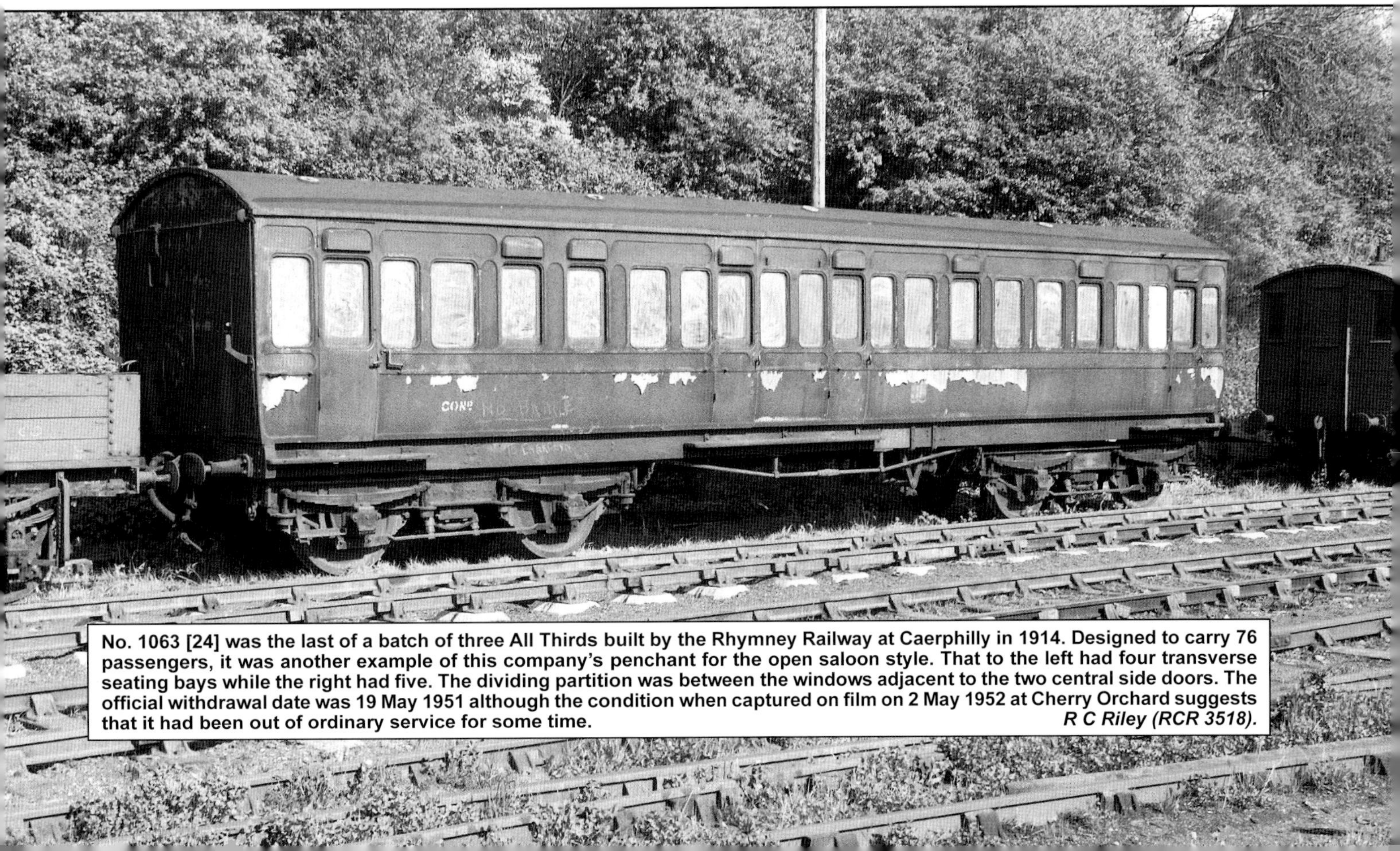

No. 1063 [24] was the last of a batch of three All Thirds built by the Rhymney Railway at Caerphilly in 1914. Designed to carry 76 passengers, it was another example of this company's penchant for the open saloon style. That to the left had four transverse seating bays while the right had five. The dividing partition was between the windows adjacent to the two central side doors. The official withdrawal date was 19 May 1951 although the condition when captured on film on 2 May 1952 at Cherry Orchard suggests that it had been out of ordinary service for some time. *R C Riley (RCR 3518).*

Above: The Taff Vale Railway was a major transporter of coal from the valleys to the ports but also had a substantial suburban passenger traffic. No 2502 [346] was one of a batch of twelve All Thirds built by Ashbury Railway Carriage & Iron Co in 1902. It was a perfectly conventional 49' 3" eight compartment coach that provided seating for 80 passengers. It was found at Rhymney on 20th May 1952. It was the second to last of the batch to remain in service, being withdrawn on 21 April 1956.

R C Riley (RCR 3468).

Opposite Page Top: The Taff Vale Railway included in its fleet coaches that were a quantum jump in size from what has been seen so far in this series. No. 4023 [14] entered service in 1905 as the largest form of Steam Rail Motor owned by the company with the coach section manufactured by Brush Electrical Engineering Co and the power unit by Manning Wardle. In this form they were 70' 4" long over buffers and provided seating accommodation for 16 First and 57 Third class passengers. No. 14 was the first of the last batch of three supplied to the company. Thereafter the SRM concept fell out of favour and the fleet of sixteen was progressively withdrawn between 1914 and 1922 as the power units became worn out. The coach sections were retained for use either as auto train trailer units or as conventional hauled stock. In May 1922, No. 14 was withdrawn and converted to a gangwayed Brake Composite to form part of a four-coach corridor train with ex-SRM Nos. 15 (All Third), 6 (All First) and 16 (Brake Composite).

No. 14 was renumbered 4023 in the GWR series in November 1922. In its new form, it was 61' 11" long with accommodation from the left: guard and van space with double doors; short Third class saloon with two transverse seating bays; entrance vestibule with double doors; longer Third class saloon with four seating bays; short First class saloon with two seating bays (this section appears to have been later down-rated to Third). All the side doors opened inwards which must have been inconvenient for the van section. Official seating was x16 First and x48 Third; there was no gangway connection at the van section end. How long No. 4023 worked as part of a four coach set is unknown. This coach was at Caerphilly on 6 August 1951; it was withdrawn in September 1955.

R C Riley (RCR 3399).

Opposite Page Bottom: The archive describes this train as a Taff Vale Railway motor set at Llantrissant on 12 July 1952 but unfortunately does not record the vehicle identities. Study of records indicates that the coach nearest the camera is either No 2506 [353] or 2507 [354]. In BR days the running number was moved from the panel below the driver's side window to immediately below the most distant passenger window. On magnification the number appears to be W2506. This coach was purpose-built as an Auto Driving Trailer All Third which entered service in 1907. A pair of very similar First/ Third composite auto trailers were introduced concurrently (Nos 78 & 79) but these had slightly differing window layouts.

Evidently intended for high density suburban work, TVR Nos 353 & 354 provided seating for 80 within a 64' long body. The larger saloon had seven transverse seating bays and the smaller had four. Passenger access was by the double doors towards the further end of the coach. The double doors at the far end were for a small luggage compartment. The GWR issued Diagram A35 to cover these coaches. Withdrawal dates were: No 2506 [353] – September 1955; No 2507 [354] – November 1957.

The further coach cannot be positively identified but may be either of No 2521 [78] (down-rated from Composite to All Third and concurrently renumbered from 6421) or No 6422 [79] which retained its composite status until withdrawal.

R C Riley (RCR 3749).

S.L.S.
SPECIAL

THE GREAT WESTERN TRUST (GWT) - BULLETIN NO. 6

Over its long and admired existence, the GWR as a Company employed a very substantial number of servants (up to WW1 vernacular) or employees (thereafter). What is striking to our current times however, is the similar longevity, contribution and extent of the social recreational and welfare activities that the GWR (and latterly BR(W)) created, maintained and wholeheartedly sponsored. This Bulletin addresses merely one example of this endeavour and gives us pause to ponder what our current era companies do in similar kind, if at all.

I am in awe of the seemingly innate ability of senior GWR players to create what today would be 'strap line' phrases of such direct but appropriate simplicity. Examples are 'Is it Safe?' no less than Sir Felix Pole's chosen phrase to headline the then revolutionary GWR lead in the Staff Safety Movement (another future Bulletin beckons I think) and today's focus 'The Helping Hand Fund'.

The background to that fund, is thankfully recorded in another GWR staff institution, the London Lecture & Debating Society annals, in which its BR(W) successor lecture '*Staff Welfare on the Railways*' of 20 Jan 1949 given by Charles Humphries Chief Welfare Officer on the WR, who explained amongst so much else of worthy study that it was the GWR's Social & Educational Union (S&EU) that formed this fund in 1924. I quote 'Another outstanding and particularly pleasant feature of external welfare is the Helping Hand fund which is held in high esteem by the staff generally and the officers of the region. A small, unnamed committee of five meet weekly to consider applications and render advice and assistance in a manner that will best serve the interests of the applicants. This relates to distress brought on by illness, bereavement, accidents and the like'. Over its lifetime, this Fund had clearly made a massive contribution to a quite staggering number of staff and their families benefiting from its assistance.

Those with access to the wonderfully informative *GWR Staff Magazines*, will find virtually every edition from 1924 containing aspects of the fund, covering not only the sums donated to it in cash, by staff themselves or through dedicated events, and how those funds had been used. The Centenary Edition of that journal of September 1935 (page 538) has a most informative table of the extensive donations of £59-11s-5d (about £4,000 today) made between 16th July and 12th August, and by whom. Tellingly it states that £87-9s-11d (about £5,800 today) was actually expended giving assistance and the £27 balance was, I quote 'met from amounts raised by Headquarters efforts', the nature of which may become clear in the Directors' involvement covered later. Of that total collected it included a guinea (One pound and a shilling; today worth about £65) given by Lady Milne, the wife of Sir James, the GWR General Manager, the GWR Swindon Amateur Theatrical Society £10, two CGMO Paddington Collecting boxes raised £2 11s 5d, and pleasing to me, the Didcot Branch gave a guinea. It also includes station donation boxes both at Paddington and over the system. This links to the item illustrated with this Bulletin, the modest original donation box once kept at Kingswear Station (7" x 3" x 9.5" long). Acquired by the Trust through an auction, we have it on display this year in our 'Recent Acquisitions' case in our Museum & Archive building at Didcot Railway Centre. Wonderfully pasted on it is a label signed by the Station Master recording £1-16-0d (about £70 today) had been collected in 1942.

Lest we purely focus upon GWR, the BR(W) *Western Enterprise* quarterly journal of the then Staff Association (replaced the S&EU from 1937) included an article by a retiring officer in which he expressly hoped that the excellent work of the Helping Hand Fund would continue for ever!

A final illuminating fact, was tabled by the ex BR(W) Signalman & author Adrian Vaughan who posted on our DRC Facebook that up to the end of the GWR, the Directors who didn't claim their due remuneration, had passed a Resolution by which any unclaimed sums were to be transferred to the Helping Hand Fund.

Need I say any more about our expectations today?

Peter Rance - GWT Trustee & Collection Manager.

THE GUARD'S COMPARTMENT

ISSUE 4

Over a long career working across the Western Region, **Mike Romans** held various positions managing and training signalling staff on the Rules and Regulations. He kindly sent the following details to help clarify the article on **Page 38**:

It might be worth pointing out that the full title of the regulation on the GWR was 'Section Clear but Station or Junction Blocked (Warning Arrangement)'. This remained unchanged until the first issue of BR(W) Block Signalling Regulations which took effect from 1 October 1960 when the title was reduced to 'Warning Arrangement'.

As explained in the article Signalmen were given authority to apply the Regulation through the Signal box Instructions - known as 'the footnotes' on the GWR. Thus, Drivers did not know where the warning could be applied and the only-way they found out was if their train was accepted in that manner. Although even this could be misleading as in conjunction with certain other Regulations a warning acceptance could be used at any signal box.

Possibly a Driver might notice a warning subsidiary signal at the 'box in rear of the section where it applied. These appeared in the GWR Rule Book in July 1934 and they were first shown in signal illustrations in the new1936 General Appendix. The manner in which a Driver was advised he had been accepted under the warning where there was no subsidiary signal was, alas poorly and inaccurately described in the article.

A lot depended on where the various stop signals were in relation to the 'box. If there was a Home Signal - in rear of the 'box - and a Starting etc Signal in advance of the 'box the procedure was that the approaching train would be brought nearly to a stand at the Home Signal and, after the signal had been lowered, as the train approached the 'box the Signalman would exhibit a green hand signal held steadily (a flag by day or lamp by night) which the Driver would acknowledge by sounding the whistle. If the Driver failed to acknowledge the green hand signal the Signalman was required to halt the train at the Starting etc Signal at the entrance to the section in advance. The procedure then was that having been stopped at that signal the Driver would understand (from the Rule Book) that when that signal was lowered it meant his train had been accepted under the warning and he should proceed accordingly.

If there was no signal in advance of the box the Signalman was required to stop the train at the box and verbally instruct the Driver that he had been accepted under the warning.

When a subsidiary warning signal was used it would be lowered only once the train had been brought nearly to a stand at the signal.

Pulling off a signal slowly does not convey the same message to a Driver as either a green hand-signal or being stopped at the most advanced signal and it had no official meaning. I know it was done when applying Rule 39(a) and its later equivalent but that, in my experience might be at a succession of stop signals controlled from the one 'box. The need to actually halt the train - when it was necessary - was critical in conveying the message to the Driver that his train had been accepted under the warning and also made sure the speed at which the train entered the section was greatly reduced.

Mike further commented that the signal described in the caption describing the arrival of No. 50 001 at Bodmin Road on **Page 50,** is in fact the Up Starting Signal.

'On the Western running line stop signals in rear of a signal box (except those reading from bay and terminal platforms) were Home Signals - suffixed according to the number of them and their relative positions. Those stop signals in advance of the signal box were all Starting signals - again suffixed according to the number of them etc.

Although possible change was considered at various times over the years by the early 1920s the outermost Home Signal on the GWR and WR was always the Home Signal. This applied irrespective of how many other Home Signals, if any, there were between it and the signal box.

--- o O o---

ISSUE 6

Stuart Johnson emailed in relation to the *Hungerford Through Time* article. In respect of the caption to the image on **Page 10**, I think it is unlikely that up passenger trains would have departed from the down platform. The picture at the bottom of **Page 15** shows that there was only a single ground disc at the London end of the down platform, and that the goods yard points (trailing to down trains) were not equipped with a facing point lock. Therefore, it would have been necessary for them to be clipped and padlocked if a passenger train was to run over them in the up direction. No doubt this was done for the working of the Royal train shown on **Page 11**. For regular terminating services, it would be much simpler, after running round, for the empty stock to be stabled clear of the main in the down mileage siding and shunted to the up platform when departure time approached.

WESTERN TIMES

Fred Finney commented regarding the lower photo on **Page 23** of Standard Class 5, that readers may be interested to see an almost identical shot taken on the same day by T E Williams at page 119 of *The Lost Colour Collection Vol. 4* (Irwell Press). That colour image reveals clearly the unique-to-Swindon practice of lining only the lower edge of the running plate.

The *Hawksworth Coaches: Part 2* article on **Page 28** prompted the following thoughts from **Philip Crook**.

- **Relative Short Life of Post War Hawksworth Coaching Stock:**

I feel that this took place for a number of reasons including the following:-

(i) More intensive use of front line main line stock; trains were often turned around in terminal stations, instead of being sent back to carriage sidings such as OOC after each journey.
(ii) The Beeching Report identified that a significant proportion of the coaching stock was only used for a limited number of times per year, thus a cull of less well used vehicles took place.
(iii) Line closures (and also intermediate stations on main lines) from the mid 1950's onwards on the Western Region, resulting in less coaching stock being required.
(iv) Replacement of most remaining suburban and secondary services with DMU's (and in some cases such as in the Eastern and Scottish Regions with EMU stock).
(v) The all steel Mark 1 coach body, mounted on an underframe that is able to withstand a 200 ton end load force, plus the provision of buckeye couplers on gangwayed Mark 1 vehicles (as also installed on former LNER and SR main line corridor coaching stock) had far superior damage resistant properties in the event of an accident or derailment when compared with pre-nationalisation coach designs utilising timber framed bodies, from any of the big four railway companies.

Generally after 1970 very few passenger carrying wooden bodied coaching stock vehicles, built to any pre nationalisation design remained in service. The most well known exceptions were some LNER buffet cars, some of which remained in service up to 1976, including on train services in the Western Region. Similarly the pre-war Southern Railway EMU vehicles such as those in 2-BIL and 4-COR units, with timber framed bodies, were also replaced in the early 1970's.

I know in later years that the Mark 1 design was criticised from a crash-worthiness perspective, but that was in comparison with later all integral body shell designs such as the Mark 3 coach.

You do make the relevant point on why some more relatively newer vehicles could have been considered for use in DMU sets, however, that opportunity was missed in the mid 1950's. When the three Hawksworth vehicles were converted for use in DMU sets (c1961, see below), most of the DMU fleet on the WR had already been delivered. I would also make the observation that the suburban compartment type vehicles used in WR suburban trains were not in my view particularly attractive vehicles to travel in when compared with any of the first generation DMU designs.

- **Hawksworth Composite Coaches converted for use in DMU sets:**

I have found some rather confusing information on the withdrawal of these vehicles (W7254W, W7804W, W7813W).

It is noted that in Hugh Longworth's book *BR Pre-Nationalisation Coaching Stock: Vol. 1 GWR & LNER* (2018), that all three vehicles were withdrawn in March 1968. However, in the same author's book *BR First Generation DMU's* (2011) a withdrawal date of December 1966 is given for these coaches.

For further information on the three Hawksworth composite vehicles used in the DMU sets please see also the Diesel Railcar Association website (*https://www.railcar.co.uk/topic/hawksworth-coaches*).This site indicates that the three Hawksworth vehicles were withdrawn in the first half of 1967.

These vehicles were replaced with three intermediate TCL DMU vehicles built by Metro Cammell, from the North Eastern and Scottish Regions (NE59528, NE59538, Sc59543) in early 1967. I understand that these were the first Metro Cammell DMU vehicles to be allocated to the Western Region.

--- o O o---

PREVIEW WESTERN TIMES ISSUE 8

Published December 2023

- **The Princetown Branch**
- **Alternative Freight Power**
- **Forty Pairs of Numberplates: Part 1**
- **End of Cornish Semaphore Signals**
- ***The Mew* - Ferry Across the Dart**